D0843001

Birds and Us

Birds and Us

A 12,000-Year History from Cave Art to Conservation

TIM BIRKHEAD

PRINCETON UNIVERSITY PRESS
PRINCETON AND OXFORD

The List of Illustrations on p. ix and List of Plates on p. xiii
constitute an extension of this copyright page

Published in the United States, Canada, and the Philippines by
Princeton University Press
41 William Street, Princeton, New Jersey 08540
press.princeton.edu

Original edition first published by Penguin Books Ltd, London in 2022

ISBN 978-0-691-23992-7
ISBN (e-book) 978-0-691-23994-1
Library of Congress Control Number 2022932387

Set in 13.5/16pt Garamond MT Std
Typeset by Jouve (UK), Milton Keynes
Printed in the United States of America

1 3 5 7 9 10 8 6 4 2

For my students, undergraduates and graduates,
who taught me more than I could ever have anticipated.

Contents

List of Illustrations

List of Plates

Preface

At the age of six I was tremendously lucky to have a teacher – Mr Govett – who was passionate about both birds and art. He read us Arthur Ransome's *Great Northern?* – a story of adventurous children discovering a breeding pair of vanishingly rare Great Northern Divers in the Outer Hebrides. We were then tasked with illustrating scenes from the story. Knowing that I liked birds – an interest encouraged by my father – Mr Govett asked me to paint the bird itself. I still remember the sense of pride in being selected for this special assignment. Little could he have known how this simple act of encouragement would shape my life.

That experience was reinforced by an extraordinary coincidence. Walking along a desolate beach on the east coast of Scotland a few years later, my father and I came across a Great Northern Diver standing disconsolately on the shoreline. Its crisp black-and-white breeding plumage implied it was in perfect condition, but something was wrong, for it made no attempt to move away. Standing awkwardly, as divers do, the bird looked up at us – pitifully, it seemed – with its blood-red eyes; it was a victim of oil pollution, which at that time was so common in the world's oceans. I could not believe that the bird I had painted in Mr Govett's class was here, right in front of me. There was nothing we could do for the unfortunate creature, and when I looked back as we left the shore, it was being chased into the sea by a dog.

These two events, I now realize, sowed the seeds of my

life, eliciting a passion for birds, a concern for their welfare, a taste for adventure in wild places and an appreciation of enthusiastic, knowledgeable mentors. Several years on, while in Nova Scotia, waiting for a flight into the High Arctic to study seabirds, I lay in bed at night listening to the haunting cries of Great Northern Divers (or Common Loons, as they are known in North America) echoing eerily across the waters of the nearby lakes.

My performance at school was undistinguished, but an interest in natural history – and some lucky breaks – allowed me to turn an obsession with birds into a career. This has been a calling that allowed me to recognize the many ways that people connect with birds, from those who feed pigeons in urban parks, or breed birds in aviaries in their back yards, hunt birds for food or fun, or train racing pigeons like elite athletes, to artists who observe and paint birds in evocative ways. There are myriad ways that we can *know* birds, and that knowledge, whether professional or amateur, scientific or anecdotal, provides us with a deeper understanding of nature itself. The Covid pandemic saw a surge of interest in bird-watching as a physical and psychological escape from the lockdown restrictions. What better evidence is there for the emotional benefits of the natural world, and the need to preserve what's left of it?

There's a widely used phrase in conservation biology – 'the shifting baseline' – which bemoans the fact that the current generation has no appreciation for what the natural environment looked like to previous generations, as their only reference is from their own childhood. A shifting baseline is just as applicable to our relationship with birds.

It is all too easy to imagine that our parents' generation shared the same concern about the worldwide decline in bird numbers, but they didn't. That awareness of a catastrophic decline in many bird populations became apparent only towards the end of the twentieth century. Looking a generation or two further back still, attitudes towards birds were different, with birds more often being seen as a resource, something to be exploited for meat, feathers, study skins and eggs – or simply destroyed because they interfered with human activities.

My aim in this book is to share my enthusiasm for birds and to explain the varied ways that our relationships with them have changed through time. It is a journey that spans several continents and twelve millennia, including my own journey as a bird researcher: from the cave art of our Neolithic ancestors, through the ancient Egyptians' bird-filled catacombs of the Nile valley to ancient Greece and Rome and the beginnings of a written history of birds, through the so-called Dark and Middle Ages and a fanatical obsession with falconry to the beginnings of science and a reappraisal of classical knowledge, to the Faroes and a community that for centuries has depended almost entirely on birds, and then to Darwin and the emergence of objective knowledge. We explore the Victorians' mania for specimens and the accumulation of bodies of ornithological knowledge as science gained traction. The twentieth century brought us to the beginnings of birdwatching and the field study of birds, triggering an extraordinary flowering of knowledge and empathy for them. Today there is massive, worldwide interest in birds, from the casual 'Have you heard a cuckoo yet this spring?' through to the more scientific, in which novel discoveries, such as those driven by new tracking technologies, have allowed us to see the migratory journeys of cuckoos and other birds in real time.

I have combined my passions for science, art and history to direct a spotlight on the multiple ways of engaging with birds. This historical, wide-ranging approach is the outcome of a lifetime of ornithological research and public engagement that has emerged from an enduring fascination about where our love of birds, and nature as a whole, has come from. By drawing attention to the fact that our present, largely empathetic relationship with birds may be temporary, I hope that we may be better able to protect birds into the future.

Our story starts in southern Spain, in a little-known Neolithic rock shelter in Andalusia. Our early ancestors are not especially renowned for depicting birds in either carvings or cave paintings, as those are few and far between, but here in this one shallow cave there are more bird images than all the other known caves put together. It is a place that for me marks, like no other, the genesis of our relationship with birds.

Great Northern (from Ransome, 1947)

1. Of Peculiar Interest: Neolithic Birds

How long have we been enchanted by birds?
Forever it seems.
Graeme Gibson (2005)

I have never seen anything quite like it.

Cueva del Tajo de las Figuras is hardly a cave by archaeological standards, more of a scoop in a vertical sandstone cliff, but inside, it feels as though I'm within the enfolding walls of a womb. With barely enough space to stand upright, the cave spans just four metres from one side to the other and two metres from front to back. I sit awkwardly on a smooth, upward-sloping floor polished by perhaps millions of visitors over millennia, the cave walls just centimetres from my face. It is light – the antithesis of the dark caverns I've been in elsewhere. And in the light I see flocks of birds.

This modest, shallow rock shelter is the deep cradle of Western ornithology: the birthplace of bird study. The walls here are covered with an exuberant Neolithic frieze of over 200 birds. Beautifully depicted, mainly in red ochre, but a few in white and yellow, the curved walls and roof are alive with flamingos, herons, raptors, avocets and many other species, dating back some 8,000 years.

This extraordinary abundance of birds in El Tajo was discovered, or rather announced to the world, in the early 1900s by a birdwatcher, Colonel William Willoughby Verner. The

Willoughby Verner posed and poised for action (courtesy J. Whitaker).

locals, who told Verner of the cave, assumed, as with anything else they didn't understand, that the images were the handiwork of the Moriscos – the Moors – who for seven centuries had occupied this part of Spain.

Multi-talented, Willoughby Verner was a soldier, military historian, natural historian, artist and intrepid adventurer obsessed by birds. Photographs during his heyday in the 1880s show him with a luxuriant moustache, broad-brimmed hat and an abundance of leather straps and belts and loops bearing his hunting paraphernalia, camera and field glasses.

Born in Edinburgh in 1852, Verner was the grown-up version of the archetypal Boy Scout; enthusiastic, energetic and with a keen interest in wildlife. Like many of his generation, however, Verner was a 'sportsman', shooting birds and climbing trees and cliffs to collect their eggs. At twenty-two he joined the Rifle Brigade and was posted to Gibraltar, later seeing active service in Africa in both the Nile Campaign (1884–5) and the Boer War (1899), during which Verner's galloping horse fell and landed on top of him, leaving him horribly injured with a badly broken leg and a displaced heart. Sent back to England, Verner, then aged fifty-two, retired from the army as a colonel in 1904. As soon as he was fit – and he seemed to make a good recovery, albeit with a permanently damaged leg – Verner returned to southern Spain to build himself a house and continue his ornithological pursuits in a region that abounded with bird life.

The British had ruled and maintained a garrison on 'The Rock' since the early 1700s. As a result of a trigger-happy officer apparently shooting one of the celebrated apes, a ban on killing Gibraltar's wildlife forced Verner to search further afield for his ornithological trophies. During successive

Gibraltar postings during the 1870s and 1880s, Verner spent his leave travelling on horseback through the south-western corner of Andalusia in search of birds. At this time rural Spain was a challenging and depressing place for travellers. Richard Ford, in his 1845 *Handbook for Travellers in Spain*, warned of cheap inns whose bedroom walls 'are often stained with the marks of nocturnal combat' – of bed bugs, or, as he calls them, 'French ladybirds'. As recently as the 1960s, English travellers in Franco's Spain still struggled with the lack of rural hygiene. Verner avoided the risky hospitality by camping, or by using his own modest hunting lodge near the tiny town of Tahivilla, nestled against his most profitable birding site – the vast wetland known as Laguna de la Janda. His diaries from this time are little more than lists of birds shot and nests robbed. But what lists! The abundance of birds by today's standard is utterly remarkable: huge flocks of Great and Little Bustards, Common Cranes, storks, ibises, herons, hawks, ducks and geese.

In May 1901, while riding along the desolate north-east edges of the La Janda lagoon, Verner's Spanish guide, Eduardo Villalba, pointed up to a large rocky outcrop and told him of the El Tajo rock shelter, and the 'obras de los Moros' (works of Moorish origin) it contained: images of 'stags, wolves and ibex, also of men and women and many other things besides'. Verner's recent injuries, however, prevented him from climbing up to investigate.

Later, he wrote:

> So engrossed was I in my pursuit of the birds, and so little did I know all those years that it was the custom of prehistoric man to make drawings in such localities and, above all, that any such drawings could have endured to this day in

El Tajo de las Figuras, with people unknown, in the early 1900s (from Molina, 1913).

such open and accessible spots as were very many of the caves, that *I never sought for them!*'[1]

Very few people at that time did know about the customs of prehistoric man. In 1910, Verner was told of another cave, further north in the Serranía de Ronda, known now as Cueva de la Pileta (Cave of the Pool), with some curious marks on the walls. This time he and his Spanish travelling companions decided to take a look. Despite his gammy leg, Verner had himself roped up and lowered into the vertical, bat-infested cave.

Led by José Bullón Lobato, the man who had discovered the cave on his land, Verner was taken aback by the vast network of limestone caverns decorated with animal images and 'mysterious script and symbols on the walls of its inmost recesses where eternal darkness reigns'. Verner's published account of 'his' discovery – in the English magazine the *Saturday Review* in 1911 – caught the eye of the 'pope of prehistory' and world expert on cave art, the irascible and egotistical Abbé Henri Breuil.[2]

On 17 November 1911 Breuil wrote to Verner arranging to meet and visit the Pileta cave together the following year. During that month-long visit Breuil duly verified the ancient – Neolithic – origins of the fish, serpents, horses, ibex, bison and myriad mysterious signs and symbols painted on Pileta's walls. Impressed by Breuil's knowledge and enthusiasm, Verner told him about the paintings he had heard about, but not yet seen, at El Tajo. Breuil was dismissive, doubting that Neolithic images could have survived for so long in such an exposed sandstone site.

Breuil returned to France, and notwithstanding his scepticism, was obviously excited, for as Verner discovered when he went to examine El Tajo for himself in the autumn of 1913, the Abbé had already sent someone to photograph the cave for him. Was it, I wondered, this seemingly underhand gesture that prompted Verner to put pen to paper and prepare his own account of El Tajo for *Country Life* magazine the following year:

> [T]he dull grey and yellow walls, and also the roof of this chamber, are absolutely covered with rude drawings in dull

> red. The most conspicuous are those of stags, the largest being over 2ft in height, with many smaller ones . . . of peculiar interest are the birds, of which a variety are shown, some with webbed feet.

Breuil and Verner subsequently visited El Tajo together, where the Abbé decided that the images were indeed Neolithic. Verner continued:

> This small cave, so awkward to reach and so slippery to enter, was in all probability a 'sanctuary', or place of worship of the folk of the Stone Age, who during their repeated visits to it, wore the rock to its present highly polished state, a condition he [Breuil] assures me he has repeatedly found in his explorations of similar caves in other parts of Spain.

The paintings are of men, deer, dogs, goats, unknown beasts and hundreds of birds. Some of the images are tiny, just a few centimetres high and are so precise and lifelike they must have been painted with a fine brush.

Verner and Breuil found, as did many of the early visitors to such caves, that the images were more easily seen if dabbed with a damp sponge, not realizing that by rendering them more visible in the short term they were helping to destroy their long-term viability. Wetting encouraged the formation of an obscuring layer of calcite and, indeed, when I visited the cave with Cadiz University colleague Dr María Lazarich in the spring of 2019, many of the images were hard to see.

Fortunately, both Verner and Breuil were competent artists and had independently taken care to trace, photograph and reproduce all the images they saw, so we have a remarkably good representation of what those birds and other animals looked like before they began to fade. I showed María a copy of the coloured painting Breuil had made of El Tajo's walls and asked her what she thought: 'Genio,' she said simply. It was true; Breuil was extremely talented, not just as an artist but in much else besides. María added for good measure that she suspected that he, and possibly Verner too, were spies for their respective countries.

Despite its ornithological significance, El Tajo remains little known, especially compared with Pileta, Altamira or the Dordogne caves, and like many of those, it is not open to the public. Its closure was nothing to do with protecting the paintings, but rather the result of a mother suing the local authorities after her daughter fell while trying to climb to the entrance in 2010. After closing the cave to the public, the local Junta installed a metal ladder to provide safe access for researchers. Looking at the vertical rock wall beneath the entrance to El Tajo, I wondered how Verner with his bad leg and Breuil with his hunched back (he had scoliosis – curvature of the spine) could have climbed up to the entrance, but a photograph in Breuil's account shows them using a ridge pole to gain access.

Both before and after Breuil's comprehensive account of the cave was published in 1929, El Tajo attracted a number of scholars of which María Lazarich and her colleagues are the most recent. Her specific objective has been to identify the birds and understand their significance.[3]

There are no fewer than 208 birds on the walls of the cave. Some 150 have been identified as comprising at least sixteen different species. The truly remarkable thing about these birds is the realism with which they are depicted. They remind me of those on the endpapers of my early Peterson field guide that all birders of my generation will know, but in

Some of the birds in El Tajo de las Figuras as drawn by Abbé Breuil (from Breuil and Burkitt, 1929).

red ochre rather than black, for the El Tajo images are simple, exquisitely drawn silhouettes of birds in profile. Most abundant are the Great Bustards, including a spectacular pageant of adult bustards and several tiny chicks.

It is this part of the cave wall that intrigues me most: several convoys of bustards in single file, one above the other, and in some instances painted on the flat surfaces between protruding ridges of harder rock, as though walking along a sunken path. It seems clear – inasmuch as anything in cave art is clear – that the artist cared only about the conformation of the birds' head and body, for it is these features that shout 'bustard'. The legs, on the other hand, are clunky, stumpy and perfunctory, drawn with two ochre-dipped fingers in a single stroke.

Of all the birds the El Tajo hunters pursued, the male Great Bustard must have been the supreme prize, for at 13kg this is among the heaviest of all flying birds. Females weigh considerably less (5 kg) – a difference between the sexes that is a consequence of their polygynous mating system. The fact that bustards feature so prominently on the El Tajo walls means that they must have been hunted and eaten, but how the Neolithic hunters caught them on the vast open plains is a mystery.

Across the world, different species of bustards have been considered the 'grandest and most majestic of gamebirds', and many different ways of catching them have been devised, giving us a clue as to how the El Tajo bustards might have been taken. During the hot, dry Andalusian summer, when the birds are desperate for water, one or more archers may have concealed themselves before dawn beside a pool or stream:

> As day begins to dawn, the bustard will take flight in the direction of the [water], alighting at a point some few hundred yards distant. They satisfy themselves that no enemy is about, and then, with cautious, stately step, make for their morning draught . . . as they lower their heads to drink . . . there is no escape.[4]

Another method might – as in India – have involved driving the birds slowly on foot towards an artificial hedge fitted with nooses, where in trying to pass the birds snared themselves.

Verner tells us both the Great Bustard and Little Bustard were abundant in the La Janda area. The former lived in groups consisting of a few old males and several females, or the females with their offspring, just as depicted on El Tajo's walls. Of the Little Bustard he says: 'It is a common sight to see flocks of these birds, varying from a few dozen to over a

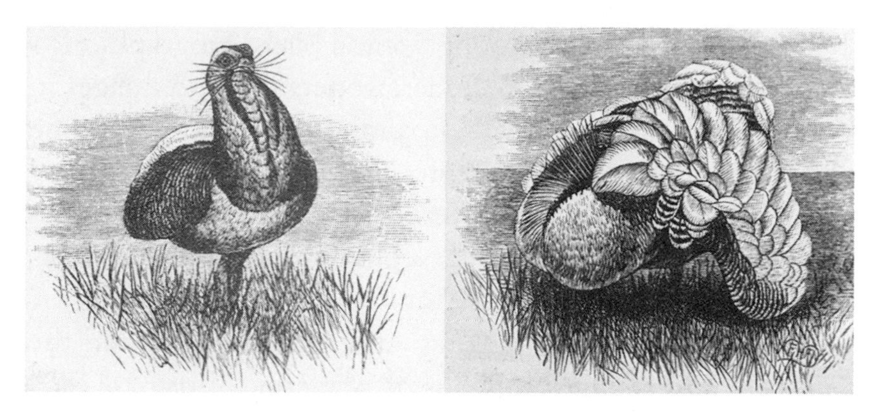

A male Great Bustard in different display postures (from Newton, 1869).

hundred manoeuvring high in the air.' And, as he could attest, both species could provide a substantial, palatable meal.

Also on El Tajo's walls are Purple Gallinules with their distinctive long toes, various herons, Cattle Egrets that seem in some cases to be poking around among the deer, a spoonbill, a pair of copulating Common Cranes near a nest with three eggs, a cluster of Greater Flamingos with their characteristic down-curved necks and bent beaks, Glossy Ibis, an avocet, eagles or vultures and ducks – all in a carefully depicted panorama. Most of the birds are represented with uncanny realism, but some are schematic – birds in flight on curved wings, like the squished 'm' of gulls drawn in childhood. And squeezed carefully among the birds and the deer there are mysterious symbols that might be the sun or stars.

There are a few people too: an archer with a bow, a man with an axe in one hand and a bird in the other. There is also a man with a bird-shaped headdress uncannily like the bird hunters of the Indus valley today, who attach a dead heron to their head, allowing them to wade within killing range without disturbing their prey.

Two features – one topographical, one geographical – provide important clues to understanding the bird images in El Tajo. The first is the cave's proximity to what was once one of Spain's great wetlands, Laguna de la Janda – a vast expanse of marsh and open water covering fifty square kilometres in the wet winter months in both Neolithic times and in Verner's day. The second is the cave's proximity to the Straits of Gibraltar – the synapse linking Africa and Europe and the gateway through which millions of migrating birds pass each autumn and spring. The Neolithic people living in this part of Spain enjoyed a massive seasonal abundance of birds.

Laguna de la Janda was drained in the 1960s to create agricultural land. The area is still a birdwatching hotspot, with a wealth of wildfowl, herons, egrets and Glossy Ibis, but a mere shadow of what it must have been previously. The last Great Bustard here died alone in 2006 after several years of failing to attract a partner.[5]

What were the Neolithic people who created the El Tajo bird images like? We know that they used stone tools, made pottery vessels, owned dogs and some domestic animals, and had a long-standing funerary culture, burying their dead in special tombs with artefacts to speed them on their way to the next life. Below the cave towards what would once have been the lake shore there are remains of houses on the level ground, where Neolithic people may have lived, but I suspect that those who visited the La Janda wetland were essentially nomadic and arrived to coincide with, and take advantage of, the great seasonal influx of birds. As is clear from their paintings on the cave walls, these people carried axes and used bows and arrows to hunt. They almost certainly used throwing sticks to bring down birds in flight, and they probably also made snares and possibly nets to trap birds in the marshes and surrounding plains. But why did they bother to paint these images of birds, deer and other animals?

When the bison paintings in the Altamira cave in northern Spain were first discovered in the mid-1800s, few could imagine that mere 'cavemen' were capable of creating such astonishing works. Instead, it was widely assumed that the paintings were relatively recent, and deliberately designed to mislead. Over the following decades, as more and more cave paintings were discovered across Europe, there was a growing realization that they were truly ancient and people like Abbé Breuil began to take note and ask what, when, how and why?[6]

The 'what' meant describing the art: animals, humans and geometric shapes; the 'when' involved dating the images; and the 'how' is concerned with the way the images were created – usually iron-ore pigments mixed with fat or blood applied with the fingers or sometimes a simple brush. None of these questions, however, is as problematical as the final one: 'why?' Over the years, this has tested the patience and ingenuity of archaeologists almost to the limit.

The way archaeologists have tried to address the 'why' – the purpose – of cave art provides an intriguing comparison with the methods employed by biologists to interpret the behaviours of birds and other animals. The parallels are considerable. Imagine an ornithologist seeing a bird adopt a particular posture, or perform a particular display (such as an aerial murmuration). They then ask themselves: Why did they do that? What is the purpose of that behaviour? In exactly the same way, an archaeologist such as Breuil, looking at the bird paintings in El Tajo or the bison and horses in

Lascaux, would have asked: What was the artist's purpose in creating these images?

The explanations offered by archaeologists for the existence of rock art have – like the colours of a chameleon – changed over time. The first idea, from the late 1800s and early 1900s, was that cave paintings were no more than art for art's sake: decorative and devoid of meaning and reflecting the simple minds of the simple people who created them. Once it was recognized that Palaeolithic and Neolithic people were not the artless, noble savages we once thought, and indeed were cognitively similar to ourselves, this argument was no longer tenable. It was replaced in the early twentieth century by 'totemism' – the veneration of particular animals and crediting them with certain powers. This idea quickly succumbed when it was realized that instead of being dedicated exclusively to particular species, as totemism predicted, most painted caves contained a mixture of animal images.

Totemism in turn was replaced by the idea of 'sympathetic magic', whose roots lay in the early years of the twentieth century. Also known as 'hunting magic', this was the simultaneously utilitarian and magical notion that creating or possessing an image of an animal imbued one with power over it, especially with respect to hunting. This is what Abbé Breuil believed, and were it true, it might represent the beginning of 'man's domination of nature', an idea that later would become a core part of Christian belief.

The blow that knocked the spell out of 'sympathetic magic' in the mid-twentieth century was the lack of concordance between the animals illustrated on cave walls and what – based on archaeological remains – the occupants of those caves actually ate.

Once the magic was dispelled, it was replaced in the late

1980s by 'shamanism'. This was an idea informed and elevated by the discipline of neurophysiology, and proposed that cave images were created by shamans whose objective was to use self-induced hallucinations to deal with supernatural phenomena, such as drought (by summoning rain) or sickness (by seeking cures). This hypothesis was devised mainly to explain the otherwise unintelligible geometric shapes that decorate many caves, including Pileta and, to a lesser extent, El Tajo. Such patterns, it was argued, were visual representations of 'entoptic' phenomena – that is, visual effects from within the eye itself – the kinds of things you see during a migraine and – I am told – during altered states of consciousness as a result of taking hallucinogenic drugs. Is it a coincidence that the timing of the shamanistic hypothesis – the 1980s – coincided with an increasingly open drugs culture?

The cave-art aficionado Jean Clottes argues that the similarity of Palaeolithic cave art over much of the northern hemisphere is in line with a widespread shamanism or religiosity. He asserts also that the extraordinary aesthetics and remarkable naturalism of many cave images is consistent with them having been produced by a few highly trained, talented individuals, which, he says, is consistent with a shamanistic or religious interpretation.[7] The sense of awe one experiences on penetrating deep inside a cave to eventually look upon a frieze of bison, horses or woolly rhinoceroses, painted as though they are emerging from the rock itself, is the same as entering a cathedral and viewing a magnificent stained-glass window or an elaborately carved altarpiece.

I was sceptical of the entoptic idea. Not that such visual effects don't occur – we all know they do – but I found it hard to imagine that they might be the inspiration for the

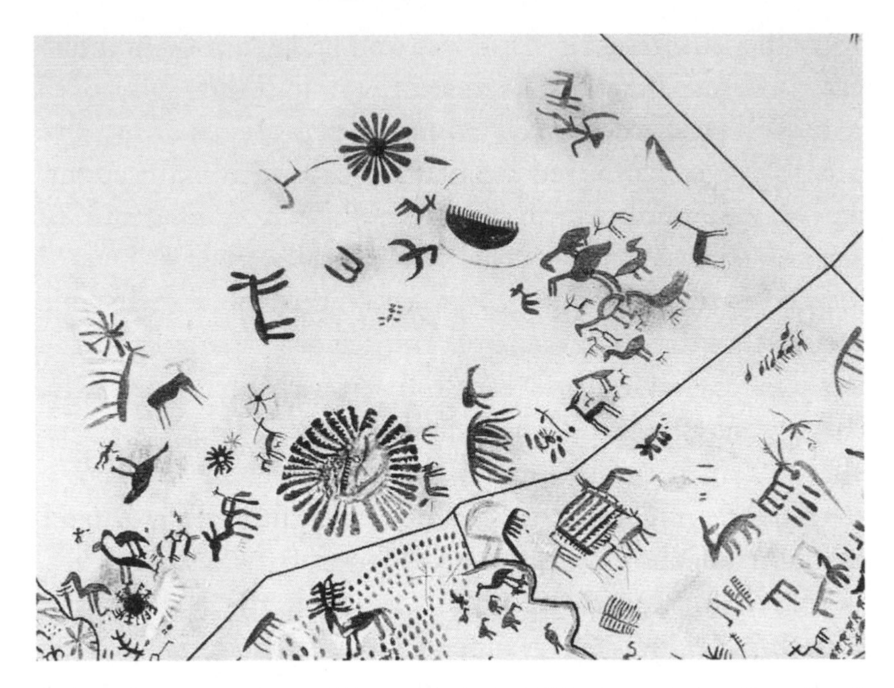

Birds and geometric (entoptic?) symbols in the El Tajo rock shelter (from Breuil and Burkitt, 1929) – the lines are Abbé Breuil's divisions of the cave surface.

geometric symbols found in caves. Then, the night after first seeing the El Tajo paintings, my sleep was a vivid shifting jumble of ochre-red images. A few days later, with the visit still reverberating in my head, I sat in the sun with closed eyes, only to see the entire array of El Tajo's images, both naturalistic and symbolic, come and go inside my eyelids against the sun. I was suddenly aware that entoptic phenomena could be very real, albeit in this case as a result of seeing the art rather than providing its inspiration. I then realized just how easy it was for a burst of euphoria to delude me into thinking I too might be a shaman.

In science, explanations for particular phenomena change over time too, but those changes are usually based on evidence. This hasn't been the case with cave art. There is increasing evidence of sorts as more and more discoveries are made, but new discoveries do little to help researchers devise rigorous tests of the various hypotheses that have been suggested to explain why cave art was created.

Some archaeologists feel that even trying to identify the purpose in cave art is pointless. I think it might be too, but it is difficult to resist. What is so striking about the El Tajo birds is their lifelike realism: they are depicted as though they are being observed, or hunted, from a distance. Compare and contrast this with seventeenth-century Dutch or Italian paintings of birds laid lifeless upon kitchen slabs. The ability of recent researchers to identify the different bird types in El Tajo is a reflection of just how well observed and rendered they were by their creators.

In some ways, we shouldn't be too surprised by this. The success of Neolithic people depended on their abilities to capture their prey – be it in the flesh or in red ochre. And, unlike most of us today, they lived much more intimately with the natural world. Apart from native hunters, only a few wildlife cameramen or field biologists spend long enough watching wildlife ever to come close to the experiences of our Neolithic ancestors. What the El Tajo artists achieved was more than mere illustration; creating those images forced others to see the birds differently. I wonder how many of today's birdwatchers would be able to create tiny images of particular birds that were instantly recognizable by others? The El Tajo painters didn't sit down and sketch a dead bird, or even a live one. Rather, they had in their mind's eye the true essence of each species that they then transferred onto the cave walls.

Possibly, just possibly, the El Tajo birds served as a field guide. Come here, climb up, carefully, into this shallow rock shelter and see what's drawn – in vibrant red, yellow and white – upon its walls. Look at them: look hard, at the different types: long necks, short necks, long legs, straight beaks, hooked beaks; see the eggs in the nest on the ground. Now, turn round and look out across the limitless wetland in front of you. Soon, in the spring, the birds will come. They'll come in their thousands, millions even, from the direction of the sea, and we will feast on them. But to hunt them effectively, you need to *know* them, not just the different types shown here, but the way they behave; the way they feed and breed in the marshes. Knowledge here is power.

Of course, bird iconography much older than El Tajo's Neolithic images exists. Both Neanderthals and 'modern' humans in the Palaeolithic exploited birds for food and for body parts – feathers, talons, beaks and bones – for ritual purposes. Palaeolithic people also created images of birds on both portable artefacts and on the walls of caves and rock shelters. But the truth is that Palaeolithic bird images and artefacts are extremely thinly spread through space and time, which is why the abundance of birds in the El Tajo rock shelter is so extraordinary.

These are the most abundant, the most diverse and the most accurately depicted birds in all cave art. They lack the grandeur of some of the better-known cave paintings, but this is partly because the mammoths, lions and aurochs and other Palaeolithic megafauna were extinct by the time the El Tajo people were painting. These wonderful bird images were created on the cusp of the transition from foraging to farming, the western endpoint of the spread of agriculture from the fertile crescent that reached as far south as the Nile valley, around 6500 BC.

Striking similarities exist between southern Spain and Egypt. Both enjoyed vast wetlands – Laguna de la Janda and the River Nile, respectively – that were wonderfully rich in migratory bird life amidst hot summers and inhospitable terrain. The crucial difference was that the annual floodwaters of the Nile poured fertility into the desert sands, allowing the ancient Egyptians to start farming and on a truly grand scale. The agricultural surpluses they generated provided the foundation of a sophisticated society with elaborate beliefs, with – as we will see in the next chapter – birds at their very heart.

2. Inside the Catacombs: The Birds of Ancient Egypt

Never again will human beings be able to rival the intense involvement with animals that was such an integral part of the ancient Egyptian civilization.

Juliet Clutton-Brock (1989)

Four million is a number I have difficulty imagining. It is eight times as many people as reside in Sheffield, the city where I live. Yet four million is the number of mummified birds – mainly Sacred Ibises – discovered in the vast subterranean catacombs at Tuna el-Gebel in Egypt in the early 1800s. Encased in ceramic jars or wooden coffins, the mummies were stacked metres high in subterranean cells carved from the soft desert rock. And it wasn't just here. Throughout the entire 1,000-kilometre length of the Nile valley from Lake Nasser in the south to Alexandria in the north, archaeologists have found over eighty animal cemeteries, thirty-one of which contain Sacred Ibises. So carefully preserved are the birds that they are still with us now, in *their* afterlife – *our* present. Unwrapped, some of the ibis mummies have the solidity of bronze sculptures, eerily reminiscent of human bog bodies.

The presence of so many bird mummies, together with numerous bird hieroglyphs and exquisite images of birds on tomb walls, reveals a society intimately engaged with birds.

Egypt's civilization was a direct consequence of the Nile.

The river's seasonal floodwaters, together with an ingenious man-made irrigation system, provided an annual dressing of fertile silt to the surrounding desert lands that resulted in prodigious agricultural productivity. The shift from foraging to farming – what archaeologists term the Agricultural Revolution – occurred some 10,000 years ago, and did so independently in different parts of the world and at slightly different times. In Egypt, the shift seems to have occurred about 8000 BC. It was also the shift from nomadism to more settled living, for the cultivation of crops such as wheat forced farmers to live near their fields. Dependency on one or two agricultural plants made farmers vulnerable to the vagaries of the weather – floods and droughts – and it was probably this that bound the Agricultural Revolution to a religious revolution: devotion to gods in exchange for abundant harvests, and sets of shared beliefs that allowed people to work together.

The abundance of agricultural produce, supplemented by the vast number of waterfowl and fish that lived in the marshes, provided the leisure, lifeblood and inspiration for the development of Egypt's elaborate culture. It was a culture in which one's presence on earth was a mere stepping-stone to the afterlife. Benevolent gods were called upon for protection; dangerous gods required appeasement – processes that demanded offerings. During his thirty-one-year reign (1187–1157 BC), the pharaoh Ramses III made offerings to the gods of no fewer than 680,714 birds – some 20,000 per year – at various temples.

During the so-called Late Period of Egyptian civilization, between 672 and 332 BC, animal gods were common, taking a variety of forms, including avian. Indeed, the distinction between human and non-human worlds was extremely vague. Life after death – eternal life – was a cornerstone of Egyptian culture, hence its preoccupation with the preservation of human bodies through mummification. The poor were buried in the desert sands and desiccated by the Sahara's dry heat. The elite were eviscerated and embalmed in an elaborate process of mummification and deposited in wooden or stone sarcophagi inside beautifully decorated burial chambers.

The Nile was as important a source of sustenance for birds as it was for people: a major migration flyway for birds moving north in spring, south in autumn. Adjacent wetlands provided food and sanctuary for many overwintering birds, just as Laguna de la Janda once did in Spain. From as long ago as 3700 BC it seems, birds have winged their way into every corner of Egyptian culture. They provided food and recreation, but their ability to fly meant that they were revered as the link between heaven and earth – between the present and the future. The mass appearance of migratory birds – ducks, geese, herons, storks, ibises and quail – at particular times of year, were emblems of new life: 'a re-enactment of the moment of creation and the conquest of death'. In a society so taken up with symbols, gods and the afterlife, it is little wonder that birds played such a vital role. As one writer commented, 'All human beings were . . . symbolically imparted with avian characteristics.'[1]

The unique cultural partnership between the ancient Egyptians and birds must have started with their Palaeolithic ancestors exploiting the seasonal influx of birds into the Nile valley as food. The transition from hunter-gatherers to the extraordinarily sophisticated civilization that we think of as 'ancient Egypt' was mediated by an abundance of food and a settled lifestyle. The Egyptians hunted birds using a variety of techniques including bows and arrows and throwing sticks, as had Palaeolithic and Neolithic people elsewhere. But they also developed much more efficient ways of catching birds.

Egyptian clap-nets were of two general types: a small version for capturing individual birds, with a central perch-trigger, and a larger one for catching multiple birds simultaneously, which comprised two net panels laid flat on the ground and held in position by ropes under tension. On pulling a release rope, the two panels clapped instantly together through 180 degrees, imprisoning any birds that were between them. With birds lured into the trapping area by bait or through the use of decoys, the clap-net was a contrivance of devastating efficiency.

Decoy birds were another novelty. These were tame or tethered individuals that, like the half-bird, half-beautiful-woman sirens of Greek mythology, lured their wild

An Ancient Egyptian clap-net, ready to be pulled and to trap a flock of ducks (from Wilkinson, 1878).

counterparts to their doom. In Egypt decoys were herons, bitterns or geese tied to the prow of a boat, whose presence as it made its way through the marshes attracted or reassured the wild birds, drawing them within killing range.

The remarkable thing about a heron decoy is that it reassures not just its own kind but also other species like ducks and geese. When I first heard of this I was suspicious, but its effectiveness is verified by several others, including the author of a seventeenth-century book on hunting: 'a fowler who wishes to obtain a multitude of waterfowl with a net need only utilize a decoy heron'; and indeed, this method was still in use in the Nile Delta in the 1930s.[2]

We know all this because of the Egyptians' extraordinary artworks: tomb paintings, engravings, sculptures and hieroglyphs. The sheer abundance and diversity of birds that the Egyptians illustrated is extraordinary: they were obsessed by them.

Among the multitude of images, one of the most beautiful is from the tomb-chapel of Nebamun, known as 'Fowling in the Marshes'. Nebamun was an elite official – a grain accountant – at the temple at Karnak who died around 1350 BC. Wheat was the single most important agricultural commodity in the Nile valley, so it is hardly surprising that Nebamun received a special burial and is the proud focus of this scene. He stands, legs astride, on a small boat, holding some decoy birds – possibly egrets – in one hand and a snake-headed throwing-stick in the other, amidst a multitude of exquisitely drawn birds, butterflies and fish. Also in the painting are his beautiful wife and young daughter. The birds include an immature African Finfoot, various ducks, herons, egrets, geese and a wagtail. At one level you could view this painting as a simple record of a happy hunting day. At another, it has been suggested, this is Nebamun ridding the world of chaos, represented by the explosion of birds rising from the papyrus swamp around him. Most probably, the image simply depicts Nebamun's vision of his afterlife.

The tombs and tomb-chapels of the Egyptian elite were designed and decorated to inspire awe among worshippers, which even in their current decayed and often desecrated

Fowling in the marshes was a recurrent theme in Egyptian tombs from 1300 to 1400 BC. Here we see a hunter (Thutmose?) accompanied by two people presumed to be his wife (on right) and daughter (smallest figure), using a throwing stick to kill birds. Note the decoy duck on the prow of the boat (from Wilkinson, 1878).

state they still do. Entering an Egyptian tomb induces an overwhelming sense of wonder, and one feels consumed by the primordial richness of the blood-red ochre pigments that seem to trigger something deeply elemental within our brains. That the artwork in Palaeolithic and Neolithic caves generates the same sensations is no coincidence – Egyptian burial chambers are simply rectilinear caves decorated in a more ornate manner.

The Nebamun tomb was discovered in the early 1800s by Giovanni ('Yanni') d'Athanasi, a Greek-born employee of Henry Salt, British consul-general in Egypt – who, with encouragement from Joseph Banks, president of the Royal Society, had become an enthusiastic collector of Egyptian artefacts. Yanni excelled at finding previously undiscovered tombs of the long-dead Egyptian elite. At that time, archaeology (if we dare call it that) was a free-for-all, with private individuals grabbing what they could, either for their own collections or to sell to museums. Such was the frenzied rivalry over the location of sites, Yanni never revealed the whereabouts of the Nebamun tomb. In an account written in 1836, soon after its remarkable treasures became known, it was said: 'We cannot learn from what part of Egypt they came, nor in what kind of place they were found.' Extensive research has since placed the Nebamun tomb somewhere on the west bank at Thebes, present-day Luxor.[3]

In what seems like an unspeakable act of vandalism, Yanni's workmen removed the most attractive images from the Nebamun chamber, cutting into the plaster surface with knives and saws: 'Sometimes they removed the full depth of the mud plaster, but sometimes they carried away only a very thin layer, and the edges inevitably suffered cracking and disintegration.' Burial tombs themselves were cut from the soft

For those adventurers undertaking the Grand Tour and brave enough to venture into Egypt, the bird pits at Saqqara were a 'must'. The German theologian Johann Michael Vansleb, describes how, in 1672, he and his colleagues had ropes tied around their middles and were lowered down into a 'well'. At the bottom they lit their tapers, and crawling on their bellies crept into a vast catacomb. There they saw 'many large stores, full of earthen pots . . . [in which there] were embalmed birds of all kinds, every bird in its own pot . . . we also found some hens' eggs, empty, but entire without any ill smell or crack'.[7]

William Wilde, Oscar's father, a medic and amateur archaeologist visiting the mummy pits at Saqqara in the 1830s, wrote:

> All was utter blackness . . . I contrived to scramble through a burrow of sand and sharp bits of pottery, frequently scraping my back against the roof . . . This continued through many windings, for upwards of a quarter of an hour, and again I was on the point of returning, as half suffocated with heat and exertion, and choked with sand, I lay panting in some gloomy corner . . . I do not think in all my travel I ever felt the same strong sensation of being in an enchanted . . . place, such as when led through the dark winding passages, and lonely vaults of this immense mausoleum.[8]

The existence of Egypt's *human* mummies has been known and written about since *c.*430 BC, when Herodotus described them. Their presumed medicinal qualities had been known (or rather, assumed) since at least Shakespeare's time, when travellers started to bring human mummies back to England. The demand for 'mummy', as this medicine was called, was greater in France than in any other country, and the ornithologist Pierre Belon, who was travelling in Egypt in the late 1540s, tells how King François I was in the 'habit of always carrying about with him a little packet containing some mummy mixed with pulverized rhubarb, ready to take upon receiving any injury from falls, or other accidents that might happen to him'. Sir Thomas Browne writing in the 1650s lamented the business of 'turning old heroes unto unworthy potions'.[9]

Interestingly, the process of human mummification occurred first – as far as we know – in Peru and Chile, about 7,000 years ago, 2,000 years before it started in Egypt. In 2019 while in Peru I saw that sarcophagi containing mummies also contained mummified birds, including toucans, but little is known about these pre-Columbian cultures, so it is unclear whether these mummified birds were food items, pets or votive offerings.

When Napoleon's army invaded Egypt in 1798, the spoils of war involved many ancient artefacts, including a large number of human and non-human mummies: birds, crocodiles, shrews, beetles and a lot of cats, from the catacombs at Saqqara and Thebes. Back in Europe, the unwrapping of mummies became a party-piece for the upper classes in both French and English salons. I saw my first unwrapped human mummy in Norwich Museum when I was five or six years old: it is an image that remains with me to this day.[10]

Napoleon's huge invading entourage included no fewer than 150 civilian 'savants', whose job was to produce the massive, multi-volume *Description de l'Égypte* (1809–29) that epitomized the link between science and empire that so characterized this era. Napoleon's trove of ancient treasure brought Egyptology into the public imagination, triggering a gold-rush for more. As one writer said: 'The long tradition

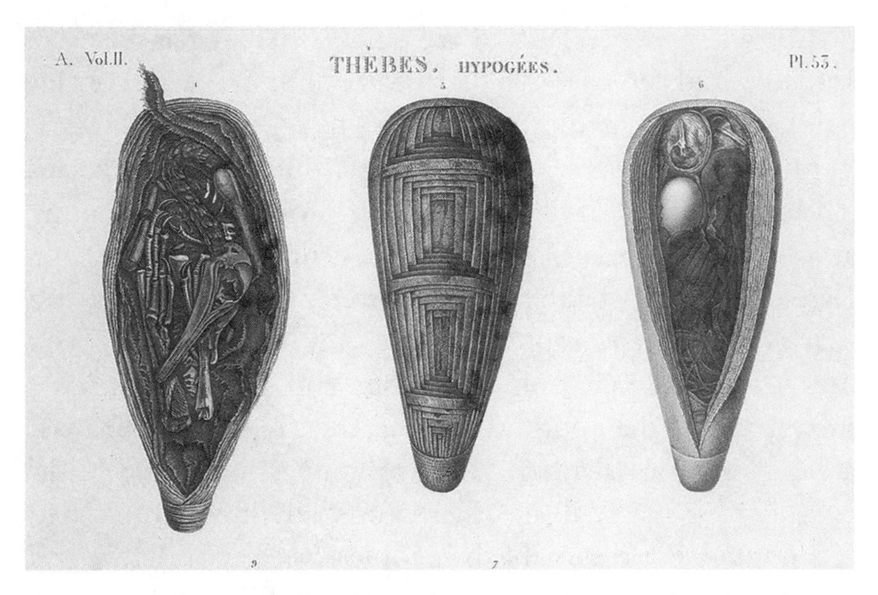

Ibis mummy discovered during Napoleon's Egyptian campaign, from the *Description de l'Égypte: Antiquités* (plates vol. II, 1812).

of travellers' accounts portrayed Egypt as the precursor of classical civilization fallen under tyrannical government, and served to justify Napoleon's imperialistic aspirations'. The British were no different, and competed with Napoleon for Egypt's riches plundered from tombs and whose acquisition they justified by comparison with a 'lost child restored to the great family of science and art, which is of no country, whose home is the world'.[11]

There was no shortage of animal mummies. A media report from 1890 described how a cargo ship arriving at Liverpool docks contained nineteen tons of cat mummies, some of which were subsequently auctioned off to museums, the rest being sold as fertilizer. At that time animal mummies were considered worthless and as late as the 1930s local people were still plundering the catacombs to use them as fertilizer.[12] Indeed, the academic study of non-human mummies did not begin until the late 1970s, since when it has become something of a niche area of research, but one with considerable public fascination.

Mummies of no fewer than ninety different bird species have been found at Tuna el-Gebel, probably representing almost all the birds likely to have occurred in that region. Ibises are the most abundant, followed by birds of prey, and many other species represented by only a handful of specimens. Interestingly, only the ibises and the raptors were mummified in their entirety; the rest were represented by just a single bone or feather. Some mummies contained birds' eggs, or nest material. In other words, anything bird-related was preserved, presumably by people taking what they found to the embalmers. As the French naturalist Bernard Lacépède said, the ancient Egyptians with their obsession for mummifying had inadvertently created a nearly complete cabinet of curiosities.

It is now thought that bird mummies served four different purposes: preserving the birds as food, as pets for deceased humans, as gods to be revered and as votive offerings. The victual mummies of Tutankhamun's funerary 'meal' comprised four different species of geese, five different ducks and a few European Turtle Doves. Birds were not common as mummified pets; cats were far more popular, as were dogs, lion cubs, gazelles and monkeys. Tomb paintings of hoopoes held in the hands of children are assumed to have been pets, although hoopoe mummies are quite scarce.[13]

The ibis was sacred and represented the god Thoth. In fact, all birds were sacred to some extent, but Thoth, the god of wisdom and writing, was special. It was apparently thought that the ibis's long, dark-tipped bill resembled a writing quill. It was also believed that ibises killed snakes; turning 'away a great plague from Egypt, when they kill and consume those flying serpents that are brought thither by the west wind out of the deserts of Libya'. In fact, ibises rarely eat snakes, and Herodotus, who related this story, may have muddled ibises with storks, who do snack on serpents.[14]

The Egyptians may have revered ibises as consumers of other pests like locusts that threatened their crops. Other birds, such as hawks, were revered for their speed, bravery or savagery, but ibises it seems were considered unclean, because of their scavenging habits, and hence were either inedible or poisonous, otherwise I'm sure they would have been on the Egyptian menu.

Where did these millions of ibises come from? Were they taken from the wild? Were they reared in captivity like so many chickens, or were some wild and some reared? It is odd that there exist tomb images of ostriches, cranes, ducks and geese being tended, herded and force-fed, but very few ibises. Does this mean that ibises were taken from the wild? Probably not, for neither are there any hunting scenes involving ibises nor ibises in clap-nets. One possibility is that their sacred state meant that ibises were excluded from tomb paintings. There is, however, good evidence for the existence of ibis 'sanctuaries'. These could have been places to which wild ibises were attracted through the existence or creation of marshy habitat and the provision of food. If they were sufficiently attractive, such sites would act as self-filling reservoirs of wild birds, from which the officiating priests could harvest what they wanted.

Another possibility is that ibis sanctuaries were places where the birds were hatched and hand-reared. The ancient Egyptians pioneered artificial incubation, possibly as early as 3000 BC but certainly by the last few centuries BC – exactly when the production of ibis mummies was at its peak. The eighteenth-century writer and savant René Antoine Ferchault de Réaumur, who was trying to develop an artificial incubation system for chickens, declared that Egypt ought to be prouder of her success with hatcheries than of her pyramids. He was right because, notwithstanding the success of the Egyptians, it took European and North American researchers until the nineteenth century to figure out how to incubate eggs artificially.

For a long time, the origin of the Egyptians' artificial incubation system was thought to be 'lost in antiquity', but since then evidence of ibis hatcheries has come to light. Excavations at Saqqara have revealed buildings 'intended for the incubation of eggs and the rearing of young birds'. Quantities of eggs were found in the excavation of the cult-buildings, and it seems possible that the breeding of Sacred Ibises (and possibly hawks, albeit in smaller numbers) was part of the cult.[15]

There is no direct evidence, but I suspect that what may have happened was that the priests running the ibis sanctuaries went out into the marshes to find and collect ibis eggs that they then brought back and artificially incubated in their 'ovens'. Once the chicks hatched they could be readily hand-reared – bird-keepers I have spoken to tell me ibis chicks are relatively easy to rear on strips of raw meat. The business of artificially hatching eggs may not be well documented for the very good reason that this was an extremely skilled task and a closely guarded secret, passed only between father and son.[16]

Another suggestion is that ibises were domesticated by the Egyptians. This is based on the flimsy evidence of a single observation by Georges Cuvier, the great French naturalist, of a mummified ibis whose 'left humerus has been broken and joined again'. His point is that 'it is probable that a wild bird whose wing had been broken would have perished before it had healed, from being unable to pursue its prey or escape from its enemies'.[17] Captive perhaps . . . but hardly evidence of domestication.

A recent study set out to test the shaky idea that the Egyptians domesticated ibises. On the assumption that populations of domesticated animals typically exhibit a genetic bottleneck – a reduction in genetic variation as a result of deliberate or accidental selective breeding – researchers painstakingly extracted the mitochondrial genes from the DNA of fourteen 2,500-year-old Sacred Ibis mummies. There was no evidence of a genetic bottleneck, and hence no evidence of ibis domestication.[18] It was worth checking, I suppose, but it always seemed much more likely that eggs of wild ibises were artificially hatched and their chicks reared in captivity.

The sheer numbers of ibis mummies show that the Sacred Ibis was once abundant in Egypt. But no more. The species now is extinct there as a breeding bird, one of the world's many birds in decline as a result of human-induced habitat loss.

The extraordinary civilization that created Giza's Great Pyramid, millions of ibis mummies and numerous deliciously decorated tombs throughout the length of the Nile valley, began to crumble about 700 BC. The Egyptian civilization was toppled by a succession of foreign invaders: the Persians in 525 BC, then the Greeks – led by Alexander the Great – in 332 BC, and finally the Romans in 30 BC. Although Greeks, and later Romans, lived in Egypt for centuries and happily incorporated many aspects of local culture into their own, they completely failed to comprehend the Egyptians' love of animal gods and despised them for it. This marked the end of the Egyptians' close bond with birds, but saw the beginning of a new era of bird-based knowledge in Greece and Rome.

3. Talking Birds: The Beginnings of Science in Greece and Rome

All birds use their tongues for the purpose of communicating with one another . . . so it seems likely that in some cases they are actually sharing information.

Aristotle, fourth century BC: *Parts of Animals*

Aristotle liked birds. He liked all forms of life, but he had a special affection for birds, probably because they are so similar to us in many ways. They walk upright on two legs, they have excellent eyesight and hearing, and their ability to sing – and in some cases to speak – sets them apart from most other organisms.

In the fourth century before the birth of Christ, Aristotle, in an ambitious intellectual feat, set out to document everything that was known about the natural world, including birds.

Educated in Plato's Academy, his main goal was to understand why things were the way they were. He was essentially the first scientist: making observations (or hearing of those made by others) and trying to make sense of them. What he learned he also taught, and his numerous books that still exist were once his lecture notes. Aristotle's abilities as a teacher are revealed in his precise and pragmatic writings and his knowledge was such that he was described as being a one-man university. His interests spanned philosophy, ethics, politics, logic and rhetoric, as well as human biology and natural history. Aristotle's pupils included Alexander of Macedon – Alexander the Great.

Of all Aristotle's bird-related information, there is one piece that stands above the rest. It relates closely to my own research on sexual reproduction. The chicken, or Domestic Fowl, as I prefer to call it, is the bird whose natural behaviour is most easily and clearly observed. Other species can be watched, but they usually offer little more than a glimpse. Farmyard fowl are tame and uninhibited, and behave much like their wild ancestor, the Red Junglefowl. The social organization of both the junglefowl and farmyard fowl consists of a dominant cockerel, a few subordinate males, and a harem of females with whom the dominant male is hell-bent on mating. Despite providing a convenient model for observing bird behaviour, the fowl is somewhat atypical in that it is polygynous, meaning that one male (the dominant cockerel) mates with the several females that comprise his harem. Most other birds, as Aristotle well knew, are socially monogamous and breed together as pairs, as most humans do.

The Domestic Fowl's domestic life – at least before large-scale commercial farming – is engrained in our minds through fairy stories and folklore. One writer aptly captured its essence: 'the cock is a jealous tyrant and the hen a prostitute'.[1]

Aristotle was fascinated by 'generation', a term that encompasses both copulation and embryo development – essentially all aspects of reproduction. Not only is generation the most fundamental aspect of animal life, in Aristotle's time and for centuries afterwards, it was also the most mysterious. It is hardly surprising, then, that Aristotle thought about it a great deal and devoted an entire book to the topic. Given the uncertainty surrounding reproductive events, his account turned out to be a mix of fact, fiction and speculation.

Domestic Fowl, partridges and small birds, he says, copulate a lot, but raptors do not copulate very much at all. His explanation is probably not one that would occur to us today, even though it anticipates some modern ideas. His suggestion is that a trade-off exists between the size of certain parts of the body and copulation frequency. The thinness or weakness of certain birds' legs, he tells us, makes them prone to copulation, adding that 'this applies also to human beings'. As he explains, this is because the nourishment that was intended for the legs is diverted into semen. Because of their short, thick legs, raptors – he says – do not copulate very frequently.

He could hardly have been more wrong.

I became fascinated by how often birds copulate after discovering that – contrary to popular belief – the females of many species, despite being paired with one male and hence socially monogamous – were actually sexually promiscuous. The consequence of such promiscuity is that the sperm from different males compete inside the female's oviduct to fertilize her eggs. My colleagues and I referred to this entire field of research as sperm competition, and to copulations outside the pair bond as extra-pair copulations. The sixty-four-thousand-dollar question was how many extra-pair copulations would it take to result in one or more extra-pair offspring?

The answer, we assumed, would depend – at least in part – on how many pair copulations the extra-pair male had to compete with. Frustratingly, no one knew, for in the 1970s when my research started, there was almost no information available on how often birds copulated – not even for farmyard fowl. How would one find out? The answer was to discover a bird species that could be monitored continuously throughout the copulation phase of its breeding cycle. The Domestic Fowl was the obvious choice, although, being polygynous, it wasn't the best starting point for trying to understand monogamy and its deviations.

I remembered that several decades earlier, in an effort to guard the nest of a pair of extremely rare Western Ospreys breeding at Loch Garten in Scotland, volunteers had undertaken round-the-clock observations, noting – among other things – the number of times the ospreys copulated. I duly obtained access to this goldmine of information and spent several weeks ploughing through years of notebooks to extract the necessary details. The results were extraordinary: 150 copulations for each clutch of two or three eggs. Although not all copulations ended in the necessary 'cloacal kiss' that signalled successful insemination, there was still an average of fifty-nine inseminations per clutch. A one-off mating with another male would not have much chance of fertilization, we guessed. Other raptors, it turned out, also had very high mating frequencies – and this is possibly the way that the male partner ensures that he is the father of the offspring he subsequently helps to rear.[2] How could Aristotle have got it so wrong? Easily. Neither he nor any of his contemporaries sat and watched undisturbed raptors at the nest in the same way as the Loch Garten volunteers had done.

On the other hand, Aristotle was right about Domestic Fowl: they certainly do copulate a lot. I was subsequently able to study the mating behaviour of free-living Domestic Fowl with Tom Pizzari, a talented PhD student, which revealed a level of copulatory sophistication that would have startled Aristotle as much as it startled us. We (Tom, mainly) showed that cockerels know and recognize each hen in their harem, remember when they have copulated with them and, most remarkably of all, adjust the number of sperm they transfer to each female depending on the time since their last mating and whether that female had copulated with another male.[3]

Aristotle may have been muddled about some aspects of reproduction, but here is the standout example of his extraordinary insight. It concerns sperm competition in Domestic Fowl and a phenomenon known to present-day biologists as 'last male sperm precedence'.

When one cockerel, Aristotle informs us, is removed from a flock and replaced by another, it is the second male that fathers most of the subsequent offspring. Not too surprising, you might think, but the significance of this – and it seems that Aristotle understood it – is that even without copulating with a second male, the hens would have continued to produce fertile eggs and chicks. This occurs because female fowl store viable sperm for up to three weeks. When a second male replaces the first and starts to inseminate the hens, his sperm take precedence – hence the term. Aristotle did not know that the female fowl's protracted period of fertility was the result of stored sperm, but he seems to have recognized that replacing one male with another produced an unexpected result – why else would he have commented on this?

A further remarkable thing is that for Aristotle (or, more likely, his informant) to have known about this, the two cockerels must have been of different genotypes – distinct breeds – such that they sired offspring with different coloured plumage. Otherwise, how could anyone have known that the second male fathered more offspring?[4]

Last male sperm precedence, we now know, occurs in many different animals, from fruit flies to finches, in which females mate with more than one male. And, of course, this is why it is worthwhile for a male to inseminate an already-mated female – there's always a chance of fertilization. For later biologists, including myself, figuring out the process by which a second male's sperm takes precedence was a fascinating challenge. Like Aristotle, I used Domestic Fowl to investigate this, but I used molecular methods – DNA fingerprinting – to assign paternity. Basically, in Domestic Fowl – and probably in most birds – the second male's sperm numerically swamps those of the first in the female's sperm stores. Simple? Yes, as are many biological phenomena when you get down to it, but demonstrating this convincingly was far from simple.

Why get excited by anything that in retrospect seems so obvious? The answer is that in other animals last male sperm precedence occurs for other reasons. In dragonflies, for example, second males physically drag the first male's sperm out of the female before introducing their own.

My research on sperm competition in birds was stimulated by discovering widespread female promiscuity and wondering about the competition between the sperm of their different partners. This became a major area of research and it would have been nice if I could tell you that it was all started by Aristotle's observations on chickens. Sadly, I cannot. The relevance of his second male sperm precedence observation lay unrecognized for millennia and was only 'discovered' by poultry biologists in the 1960s. This was a couple of decades before the subject of 'sperm competition' became popular among evolutionary biologists. Even so, I love this link – thank you, Aristotle – between the past and the present.[5]

Living on the island of Lesbos between 346 and 343 BC, Aristotle began to document what was known about the natural world – including birds – in the first truly systematic study of biology. His notes and observations of birds were far-reaching, and one of his most optimistic efforts was to try to create a classification of birds. He did so on what we can call 'functional types' or lifestyles – ways of making a living. For birds, this meant dividing them into raptors, marsh birds, water birds and so on, but this was not a taxonomic classification reflecting their true phylogenetic (evolutionary) relationships – that would have to wait for more than a millennium. Having said this, Aristotle did see an overall 'scale of perfection' in the animal world – with humans at the apex, plants and minerals at the bottom, and birds lying close to the top, just below quadrupeds and whales.[6]

More specifically, Aristotle understood that feathers were analogous to the scales of reptiles, and that there were different types – including the soft hair-like feathers of ostriches. He also knew that birds changed their feathers – and often their appearance – at certain times of year. He knew about the internal structure of birds, having dissected a dove, a duck, a goose, an owl, a pigeon, a partridge, a quail and a swan – but curiously, no passerine birds. He noted all the major organs, without much understanding of their roles, but commented on the existence of both a crop and a gizzard in some birds; on the Eurasian Wryneck's long tongue; and on the differences in the appearance of the gonads of the two sexes. He noted too, the zygodactyl feet of woodpeckers – two toes pointing forward and two back – unlike the three-forward-one-back of most other species. Aristotle is often said to have been the first to describe the development of the chick. He also thought, erroneously, that birds' eggs were laid with a soft shell – to ease their passage – that hardened on exposure to the air. A touching idea, but not true.

Aristotle amassed information, both from his own observations and from those who lived close to nature such as fishermen, beekeepers and bird-catchers. He then sought patterns within the mass of accumulated 'facts' and from these generated general explanations for what was observed. He was sufficiently objective and open-minded to realize he might often be wrong: 'But the facts are incomplete, and if at any future time they are better established then more credence should be given to the evidence . . .'[7]

His approach was one that characterized science for much of its history. All areas of study go through an observation phase – equivalent to Aristotle's amassing of information. Then follows a period of trying to make sense of those observations; attempting to find some generalities or patterns that provide general explanations. Aristotle's followers in the next 1,500 years, however, saw his explanations as final. It was as if he had opened the door to understanding, then pulled it firmly shut behind him, because nothing more was required. But of course, nothing could be further from the truth. The difference between Aristotle's 'science' and what started in the mid-1600s was that Aristotle's 'explanations' were really just ideas – ideas that required the rigorous testing and verification introduced by the Scientific Revolution.

None of this is to undermine Aristotle's remarkable achievements. His approach, which included taking information from others, meant that he was bound to make some mistakes. Some later writers, including the Nobel Laureate biologist Peter Medawar and his wife Jean, considered Aristotle to be a dud. They referred to his works as 'a strange and generally speaking rather tiresome farrago of hearsay, imperfect observation and wishful thinking'. But that's an overreaction, and places far too much emphasis on Aristotle's errors. Darwin was much more positive: 'Linnaeus and Cuvier have been my gods . . . but they were mere schoolboys to old Aristotle.' Similarly, for his recent biographer Armand Leroi, Aristotle was a scientific pioneer and the father of natural history who recognized that 'In all natural things there is something of the marvellous.'[8]

The Greeks' relationship with birds played a pivotal role in subsequent Western attitudes to nature in a way that the Egyptian attitudes did not. Aristotle considered birds special on account of their songs, calls and cries. He asks whether birds possess 'reason' – whether they have the ability to think rationally and to know what they are saying. The moral status of birds in Greek culture depended not on whether they could feel pain or experience pleasure – two traits Aristotle did not doubt – but whether they behaved rationally. Aristotle and other Greek philosophers believed that rationality was closely linked with language. The ability both to teach and to learn were, he believed, signs of a rational being. There was no speech without reason and no reason without speech. The vocalizations of birds are speech, so birds – above all other non-human animals – are rational and worthy of respect, he said.[9]

These ideas have their origin mainly in the observations of captive birds and I am continually amazed by how early in human history many fundamental ornithological insights were made. Wealthy Greeks kept birds as pets, and the Common Nightingale, with its luscious song, was a favourite. Aristotle knew – and this is now well established – that young birds acquire their song, in part at least, from hearing their parent. He was wrong in assuming it is the mother nightingale who sings and teaches her offspring – a myth that took centuries to dispel. It is the male, as in most birds, that sings and from whom the young birds learn. In terms of this discussion, however, that is irrelevant. What made birds seem rational to Aristotle was the fact that species like the nightingale have the ability to teach their offspring, and that they in turn are receptive to being taught.

The realization that birds acquire their song from a parent was probably derived from Aristotle's observation that if young birds 'have been removed from the nest and have heard other birds singing . . . some sing a different note from the parent birds'.[10] I can appreciate why this made such an impression on Aristotle. I once had a pet Eurasian Siskin that had been reared by canary foster parents, and instead of uttering the typical wheezy siskin refrain, this little bird belted out pure canary song, an incongruity that stopped me in my tracks every time I heard it.

The other birds the Greeks enjoyed and marvelled at were those that can be taught human speech: parrots, starlings and corvids such as jays, magpies and ravens. For some, the ability to mimic human speech was the pinnacle of rationality, compellingly reinforced by the similarities in the way both birds and children acquired language.

Writing several centuries after Aristotle's death, Plutarch says:

> As for starlings and crows and parrots which learn to talk and afford their teachers so malleable and imitative a vocal current to train and discipline, they seem to me to be champions and advocates of other animals in their ability to learn, instructing us in some measure that they too are endowed with rational utterance.[11]

Plutarch was inspired by the example of a pet Eurasian Jay kept in a barber's shop and able to imitate human voices, animal sounds and mechanical noises. One day a funeral procession stopped in front of the barber's shop during which time the trumpeters continued to play. After the procession had passed, the jay ceased to utter its regular vocalizations, but later produced a perfect rendition of the trumpeter's tune. Plutarch attributed the bird's temporary silence to it consciously working out how to replicate the musicians' melody.

Mustering other evidence to support his idea of the rationality of birds, Plutarch points out that species like jays and Common Starlings do not imitate sounds at random, but are very specific in what they mimic – an observation amply verified by more recent studies – suggesting to him, at least, that birds are capable of conscious thought.[12] In the same vein, Porphyry of Tyre, writing in the third century AD, was convinced that birds knew what they were saying to each other – but our inability to understand them is no different from when we hear a foreign language, which he says is analogous to the 'clangour of cranes'.[13]

Plutarch and Porphyry attributed much more rationality to birds than Aristotle ever did. They assumed that birds' reason extended to prophetic and divinatory abilities, reinforcing the view that birds were closer to God than humans. More cautious and more rational, Aristotle finally decided that the ability of certain birds to mimic the human voice was nothing more than imitation. In doing so, he rejected the idea of avian rationality, thereby helping to set the agenda for the Christian view, in which birds are distinct from us.

It was a distinction that allowed Aristotle to write:

> Plants exist for the sake of animals . . . and animals for the good of humankind – the domestic species for his use and sustenance, and most if not all the wild ones for his sustenance and for various kinds of practical help as a source of clothing and other items . . . nature has made all these . . . for the sake of humans.[14]

This was a convenient idea that later re-emerged in the Bible: 'And God said, Let us make man in our image, after our likeness: and let them have dominion over the fish of the sea, and over the fowl of the air . . .'[15]

As with all previous cultures, and many future ones, birds were a resource.

A thread runs through the Palaeolithic, Neolithic and ancient Egyptian eras into ancient Greece and Rome and beyond, connecting people's ideas and beliefs about birds. Great travellers and traders, the Greeks had been present in Egypt since at least the eighth century BC. It has been argued by some that their later civilization – art, technology, religion, burial rituals, architecture and taste for spectacular sculpture – owed much to what they saw and learned there.[16]

It is almost as if, from narrow Egyptian beginnings, Greek culture welled up and flooded across the landscape, depositing its fertile ornithological ideas across different modes of thought. By around 500 BC the Greeks' ideas about birds, or at least the way they were articulated, were becoming increasingly sophisticated and employed to better understand, influence and control the natural world. We know about these ancient thought processes because, unlike the Egyptians, the Greeks left an abundant written legacy.

The classical era spans a thousand years – from 500 BC to AD 500 – and a vast geographic empire. In terms of the study of the natural world, Aristotle was its main player in Greece, but in Rome it was Pliny the Elder. Their approaches to understanding birds and our relationship with them were as different as chalk and cheese. Separated by more than three centuries, Aristotle and Pliny are often discussed as though they were contemporaries, and even as though their expertise was similar. As a pair, they anticipate a situation that persists today: the careful, intellectually innovative scientific type – Aristotle – contrasted with the enthusiastic, sometimes careless popularizer – Pliny.

Over the centuries, the writings of both men have been hugely influential in the way we think about birds and other animals. People respected Aristotle's authority, but were inspired by Pliny's encyclopedic span and accessible style. As Aristotle's biographer Armand Leroi says: 'It was Pliny rather than Aristotle who provided the model for Renaissance natural history even if it was Aristotle, happily, who provided most of the substance.' And herein lies my fascination with them. On the one hand, I am in awe of Aristotle's brilliance and his intellectual efforts to understand the natural world. On the other, I admire Pliny's popularization of his predecessor's hard-won knowledge. The difference between them is like that between today's professional scientists who publish their findings in academic journals and the writers of popular natural-history books whose information is often an accessible digest of the scientists' efforts.

Pliny the Elder produced the most extraordinary and enduring encyclopedia of the natural world. He is thought to have died from the inhalation of toxic fumes in Stabiae in AD 79 while attempting to rescue friends by boat from the erupting Mount Vesuvius – the same eruption that smothered Pompeii and Herculaneum. Trained initially as a lawyer, he joined the army as an officer, as was typical of his elite equestrian class, but first and foremost he was a scholar. He lived in various parts of the empire, including France, Spain and North Africa, and during Nero's repressive regime he kept his head down by writing innocuously about grammar. Pliny started his vast natural history encyclopedia around AD 70, when his friend Vespasian was emperor, and completed it some seven years later. The book, among the largest from ancient Rome, and the only one of Pliny's to have survived, spanned mineralogy, geology, astronomy, botany and zoology, with information gleaned from a huge array of sources, including Aristotle.

For 1,500 years Pliny's works dominated all thoughts about the natural world. It was only once the Scientific Revolution began its reassessment of ancient knowledge in the 1600s that people began to question Pliny's authority. Increasingly thereafter, his work was viewed as:

> a repository of tales of wonder, of travellers' and sailors' yarns, and of superstitions of farmers and labourers. As such it is a very important source of information for the customs of antiquity, though as science, judged by the standards of his great predecessors, such as Aristotle . . . it is simply laughable.[17]

Eagles, Pliny says,

> lay three eggs, and generally hatch but two young ones, though occasionally as many as three have been seen. Being weary of the trouble of rearing both, they drive one of them from the nest: for just at this time the providential foresight of Nature has denied them a sufficiency of food, thereby using due precaution that the young of all the other animals should not become their prey.

There are two ideas here, one right, the other wrong. Eagles often lay more eggs than they rear, and usually one chick is killed by its older sibling, especially if food is short. This 'brood reduction', as it is known, has evolved because – strange as it might seem – it results in the eagles leaving more descendants overall throughout their lifetime. It has not evolved to minimize predation on other species as Pliny supposed.[18]

Of the Common Cuckoo, Pliny states:

> It always lays its eggs in the nest of another bird, and that of the ring-dove [Common Wood Pigeon] more especially, mostly a single egg, a thing that is the case with no other bird; sometimes, however, but very rarely, it is known to lay two. It is supposed that the reason for its thus substituting its young ones, is the fact that it is aware how greatly it is hated by all the other birds; for even the very smallest of them will attack it. Hence it is, that it thinks its own race will stand no chance of being perpetuated unless it contrives to deceive them, and for this reason builds no nest of its own.

Some of Pliny's knowledge of the cuckoo is lifted from Aristotle, who was the first to document this species' parasitical breeding habits. But Pliny's account is a mix of fact and fiction. Cuckoos usually deposit only a single egg in each host's nest, although occasionally a second cuckoo adds an egg. The Common Wood Pigeon, however, is rarely parasitized. Cuckoos are indeed attacked by other birds, in their attempts to avoid being parasitized, but being *hated* by other birds is not the reason cuckoos are parasitic. Brood parasitism evolved because brood parasitism works.[19]

Pliny adds:

> In the meantime, the female bird [the host], sitting on her nest, is rearing a supposititious and spurious progeny; while the young cuckoo, which is naturally craving and greedy, snatches away all the food from the other young ones, and by so doing grows plump and sleek, and quite gains the affections of his foster-mother; who takes a great pleasure in his fine appearance, and is quite surprised that she has become the mother of so handsome an offspring. In comparison with him, she discards her own young as so many strangers, until at last, when the young cuckoo is now able to take the wing, he finishes by devouring her.

Yes, the foster parents lavish care onto their uninvited guest as though it was their own offspring, but the foster parents do not discard or abandon their own young; Pliny has overlooked Aristotle's accurate observation that the newly hatched cuckoo chick ejects the host young (documented in detail by Edward Jenner in the 1780s – but even then not universally accepted). Nor does the young cuckoo, as Pliny states, devour its foster parent – a suggestion based on the fact that when feeding its enormous chick, the foster parent often puts its entire head inside its mouth.[20]

On Indian Peafowl, Pliny tells us when the male 'hears itself praised, this bird spreads out its gorgeous colours, and especially if the sun happens to be shining at the time, because then they are seen in all their radiance, and to better advantage'.

A peacock's inclination to display has – obviously – nothing to do with hearing itself praised, but research in 2013 showed that sunshine is important, with males specifically orientating themselves about 45 degrees to the right of the sun's azimuth, with the female positioned directly in front. Oriented in this way, the male shows off the array of iridescent eyespots on his tail to their greatest effect.[21]

To his credit, Pliny says that the idea that peacocks are both vain and spiteful 'in the just the same way that a goose is "bashful", appears to me to be utterly unfounded'. Similarly, he debunks the idea that 'at the moment of a swan's death, it gives utterance to a mournful song' – swan song. He says: 'This is an error, in my opinion, at least I have tested the truth of the story on several occasions.'[22]

In describing the song of the Common Nightingale, Pliny says, 'That there may remain no doubt that there is a certain degree of art in its performances, we may here remark that every bird has a number of notes peculiar to itself; for they do not, all of them, have the same, but each, certain melodies of its own.' Absolutely correct. He adds an interesting comment saying: 'Men . . . have been found who could imitate its note with such exactness, that it would be impossible to tell the difference.' Such is the extraordinary quality of the nightingale's song, imitating it with exactness is extremely difficult, but this is what happened to save a live radio duet that was supposed to occur between the cellist Beatrice Harrison and a nightingale in her Surrey garden in 1924. The bird, which had accompanied Harrison on previous nights, declined to perform in the presence of the recording equipment. At the last moment, and unbeknown to the BBC's listeners, a 'siffleur' – probably

Madame Maude Gould, also known as Madame Saberon – launched into a coloratura performance to save the day.[23]

Of the partridge:

> In no other animal is there any such susceptibility in the sexual feelings; if the female only stands opposite to the male, while the wind is blowing from that direction, she will become impregnated; and during this time she is in a state of the greatest excitement, the beak being wide open and the tongue thrust out. The female will conceive also from the action of the air, as the male flies above her, and very often from only hearing his voice.

Sadly, not true.

> Caprimulgus is the name of a bird [the European Nightjar], which is to all appearance a large blackbird; it thieves by night, as it cannot see during the day. It enters the folds of the shepherds, and makes straight for the udder of the she-goat, to suck the milk. Through the injury thus inflicted the udder shrivels away, and the goat that has been thus deprived of its milk, is afflicted with incipient blindness.

A fabulous myth that gave rise to the bird's common name of 'goatsucker', still familiar to birders today.

As Pliny informs us, the Romans ate birds of all kinds. They differed from their predecessors, however, by feasting on avian novelty, eating the entrails, brains, testicles, gizzards and tongues of unusual birds. As in certain parts of the world today, bizarre food items for the Romans signalled exclusivity and status – and to many of us now, a kind of depravity.

Romans ate thrushes, presumably winter migrants, captured and then fattened in special aviaries. Some at least were allowed to fly out from a roasted wild boar as it was cut open at the table – a forerunner of the four and twenty blackbirds baked in a pie. The tongues of birds were popular among Roman epicures and no more so than those of the nightingale. It seems unlikely that these really were tongues that were eaten. A nightingale's tongue, like that of most small birds, is barely worth the effort, consisting of little more than two barely digestible hyoid bones and a few meagre scraps of muscle. Later, as Mrs Beeton's famous nineteenth-century cookbook makes clear, what were commonly referred to as larks' 'tongues' were actually their breast muscles, which were much more substantial and tasty. It is unlikely that elite Roman diners knew or cared about the difference between a lark's tongue and its breast muscles. Parrot tongues were also a Roman favourite, and in this case fairly substantial, for these birds possess a large fleshy tongue that they use both for manipulating food and for vocalizing, just as we do.

The ultimate tongue in Roman cuisine, however, was that of the flamingo. There were no flamingos in Italy, so the birds – Greater Flamingos – must have been imported from elsewhere, probably Spain, southern France and North Africa. As later anatomists demonstrated, the flamingo's large, erectile, turgid tongue has evolved to pump water through the bird's beak, so that any edible particles, such as diatoms, seeds and tiny brine shrimp are trapped on the 'lamellae', in much the same way as krill are filtered from seawater by baleen whales.[24]

The flamingo features in *Alice's Adventures in Wonderland* precisely because it is such a surreal and spectacular bird. Its scientific name *Phoenicopterus* – meaning crimson wing – refers, as does 'flamingo' itself, to the bird's flame-red plumage, whose colour derives from the carotenoids in its diet. With their pigmented plumage, long neck, long legs and curiously constructed heads, one can imagine flamingos featuring on Roman dining tables as lifelike mounts, much as peacocks and swans did at medieval banquets. The flamingo's habit of feeding its chick on a crimson brine-shrimp soup dribbled from the bill into the mouth of its offspring almost certainly gave rise to the myth of another water bird, the pelican, feeding its young on blood pierced from its own breast. The great French naturalist and author the Comte de Buffon, in his vast animal encyclopaedia of the late 1700s, mentions how the flamingo was held in such high esteem by the Romans that:

> When Caligula had reached such a pitch of folly as to fancy himself a divinity, he chose the flamingo . . . as the most exquisite victim to be offered up to his godship; and the day before he was massacred, says Suetonius, he was besprinkled at a sacrifice with the blood of a flamingo.[25]

The first-century cookbook by Apicius, whom Pliny the Elder described as 'the most insatiable gorger of all gluttons', includes a recipe for flamingo and is responsible for establishing the idea that the flamingo's tongue was the ultimate in Roman gourmandizing. Emperor Vitellius, renowned for his cruelty and gluttony (and who was assassinated in AD 69 after just eight months in office), was once served a feast comprising 2,000 fishes and 7,000 birds. At one of his own banquets, Vitellius presented his guests with a platter comprising the livers of pike, the brains of peacocks and pheasants, the milt of lampreys and the tongues of flamingos. Another emperor, Heliogabalus (204–22), was said to have served up 'dishes filled with the tongues of flamingos'.[26]

I have always found the idea of eating tongue repulsive. When I was a child my family used to serve tinned cow's tongue as a Christmas 'treat', but I studiously avoided it. My reluctance is illogical, since a tongue is merely muscle just like other parts of animals that are eaten routinely, but there's something too intimate about eating an animal's tongue. For the same reason, I have never been tempted by a pig's pizzle, or *huevos de toro* or other offally bits that were regularly eaten in the past. A flamingo's tongue seems similarly unattractive, but, according to my anatomist colleagues, it is both muscular and fatty, and, once the recurved spines are removed, is probably good eating. Buffon noted that 'some of our navigators, whether from the prejudice derived from antiquity, or from their own experience, commend the delicacy of that morsel'. Buffon's navigators include the natural historian Jean-Baptiste Du Tertre, who visited the Caribbean in the mid-1600s and who said of the American Flamingo: 'their tongue is very large, and near the root there is a lump of fat, which makes an excellent morsel'. The pirate naturalist William Dampier shot and ate Greater Flamingos in the Cape Verde Islands in 1683, reporting that their flesh was 'lean, black and savoury, with the tongue being particular tasty and a dish for the king's table'. In the nineteenth century another naturalist, Alcide Charles d'Orbigny, a disciple of Cuvier and one of Darwin's many correspondents, commented that he saw a lake in Egypt 'covered in small boats going out to hunt flamingos. These boats would return full of birds, from which the Arabs removed the tongue, in order to extract from it, by pressure, a greasy substance that they used as fat'. Not everyone shared the Romans' lingual enthusiasm and the sportsman Abel Chapman, author of *Unexplored Spain*, in

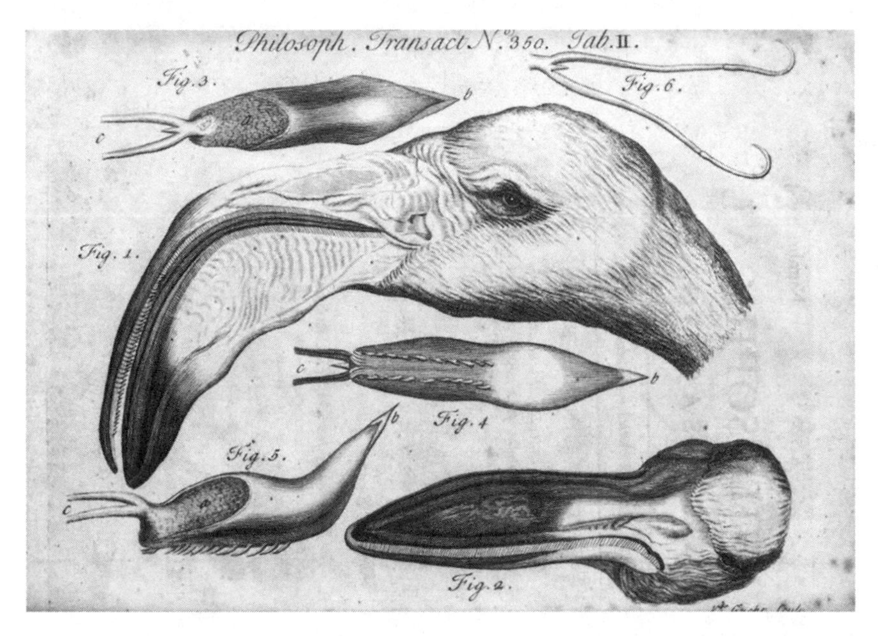

The head and tongue of a Greater Flamingo as drawn and described by James Douglas in the Royal Society's *Philosophical Transactions* (Douglas, 1714).

1910, found them: 'Quite uneatable – tough as India-rubber; even our dogs refused the delicacy.'[27]

I was intrigued by all of this and determined to see and taste this delicacy for myself. My inquiries were rewarded by being sent the head of a Greater Flamingo, sadly inedible because it was preserved in industrial alcohol, but dissecting it was a revelation. The tongue was surprisingly fleshy, and very, very fatty; embedded in it were some elongated cartilage structures – extensions of the hyoid tongue bones – from which a gourmet would have to suck the fat. And then another lead: an Italian ornithologist wrote to say how he had once cooked a flamingo using a modern take on Apicius' original recipe. He added: 'Should you happen to pass through northern Italy we can prepare and taste a couple of tongues together, trying to stick to Apicius' advices. Just tell

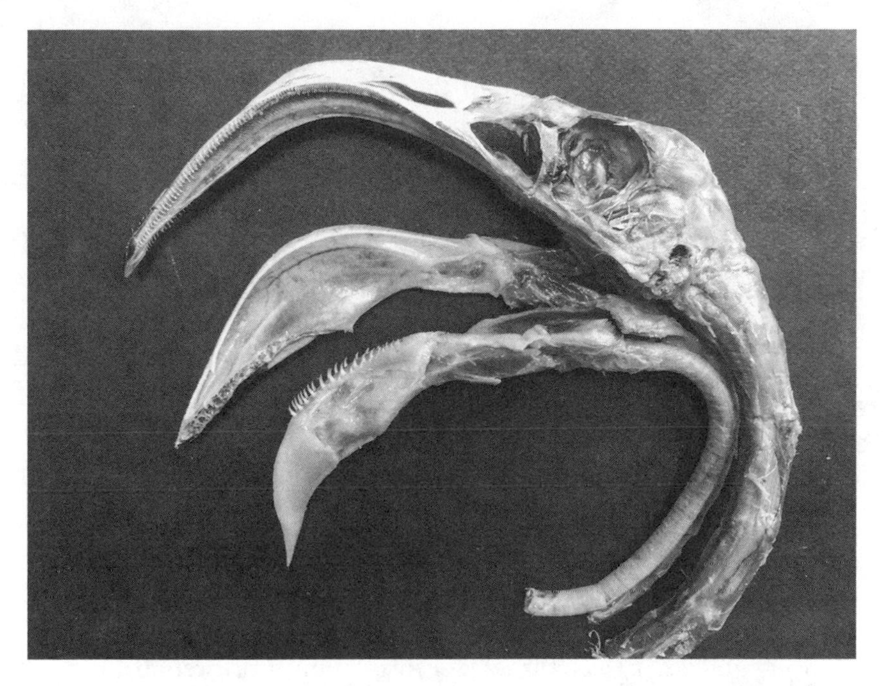

My dissection of the head of a Greater Flamingo, showing, from top to bottom: the upper mandible, lower mandible and the tongue that in life would lie inside the lower mandible (photo: Tim Birkhead).

me with some advance, so that I can alert my taxidermist colleague not to throw the tongues away when casualties from crashing into wires will be available'. Tempting.[28]

While it is true that Pliny's writings on birds helped to perpetuate many ancient myths and much erroneous information, it may be slightly unfair to pit him against Aristotle in the way I've done here. Aristotle's information about birds comes from his lecture notes, or possibly from those of his students, hence their dry, concise nature. What we do not have are Aristotle's writings intended for a general readership, which were lost many centuries ago. Imagine if they, rather than his lecture notes, had survived. Would our opinion of Aristotle be more similar to the way we think of Pliny?

Perhaps not, for it is unlikely he'd have ever compromised on what he felt was the truth. And as the Roman man of letters Cicero says, Aristotle's popular accounts were as beautifully written, as 'a river of gold'.[29]

The myths about birds with which both Aristotle's and Pliny's works abound have their origins in the past, and notably in the fifth century BC and Aristophanes' comedy *The Birds*. It is here that we discover the true value of birds:

> You don't start on anything without first consulting the birds,/ whether it's about business affairs, making a living, or getting married./ Every prophecy that involves a decision you classify as a bird./ To you, a significant remark is a bird; you call a sneeze a bird,/ a chance meeting is a bird, a sound, a servant, or a donkey – all birds/ So clearly, we are your gods of prophecy.[30]

The key to understanding this otherwise puzzling passage is the fact that in its original Greek, the word for bird, *ornis*, is also the word for an omen. In other words, because birds are augurs they were to be consulted whenever a decision was to be made.

The seasonal appearance of migratory birds like Barn Swallows or White Storks in Greece was associated with the coming spring, and the idea that this was a propitious time for farmers to plant crops. The drumming of a woodpecker sounds like distant thunder or the drumming of rain on a roof, hence their supposed ability to anticipate wet weather. The fact that Carrion Crows and Northern Ravens typically operate as pairs throughout the year gives a sense of fidelity – which turns out

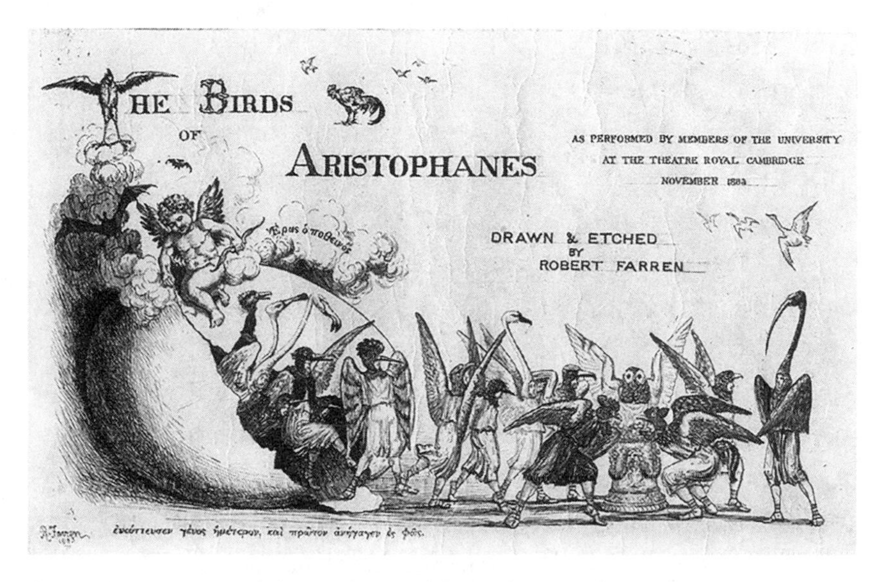

Artist's impression of Aristophanes' *The Birds*, one of a series of etchings linked to a performance of the play by King's College, Cambridge students in 1883 (courtesy of King's College, Cambridge).

to be true, in terms of not only their enduring pair bond, but also their sexual fidelity. Another association is that between ravens and death. Crows, ravens and vultures were a frequent sight in the aftermath of war, scavenging on corpses, so the link between these birds and death is hardly surprising. The Greeks also knew that ravens are smart, both from birds kept as pets and from the fact that they seemed somehow to know when a battle had occurred and turned up to feast on the dead. We now know that birds have a better sense of smell than we once supposed, so this may be one way they discover new food sources, but ravens also take their cues from other birds, following individuals that seem to know what they are doing, or look smugly well fed.

Since the earliest written records, parallels have been drawn between birds and people: as greedy as a gannet, daft as a coot, a cuckold and the foolish guillemot. Many of these attributions date back centuries, and the fact that they have survived intact for so long is probably an indication of their appropriateness.

Pliny devotes more space in his *Natural History* to eagles than almost any other group of birds, telling us that they are the most honourable and strongest of birds. As we might expect, there's plenty of eagle misinformation, including the eagles whose nests contain a stone within a stone with great medicinal powers. Pliny repeats Aristotle's accurate observation that eagles need a 'large tract of country to hunt over' and how in the early hours of the day eagles perch 'quite idle' and fly mainly in the afternoon. He also describes how in 104 BC the eagle's reputation for honour and strength resulted in it being adopted by the Roman military as their standard (*aquila*) when it became the custom to carry the eagle into battle.[31]

Ferocious animals have been used as visual metaphors in various cultures, but eagles were pre-eminent in terms of their size, visual acuity, effortless flight and, above all, the killing power of their massive feet and talons. Eagle motifs are the emblems of numerous nations. The Egyptians' more modest veneration of falcons was part of the same tradition. Through vast tracts of time, birds of prey have been among the most enduring symbols of supremacy and, as we will see in the next chapter, eventually came to define an entire social class.

4. Manly Pursuits: Hunting and Conspicuous Consumption

He cannot be a gentleman that loveth not hawking and hunting.

James Cleland (1607)

The last Anglo-Saxon king of England, Harold Godwinson, who died at Hastings in 1066 allegedly with an arrow in his eye, was said to be obsessed with falconry, only ever putting his bird down when needing both hands to eat.[1]

The Bayeux Tapestry opens in the year 1064 as Harold sets out on horseback – with a hawk on his left fist – on a diplomatic mission to meet William (the Bastard), Duke of Normandy. Trotting southwards towards the Channel, Harold's aim is to leave for France from Bosham on the West Sussex coast. Once at sea, however, his boat is blown off course and, making landfall at Ponthieu, he is captured by Guy (aka Wido) of Ponthieu, where – hawk still on hand – he is taken under guard to Guy's palace at Beaurain. From there Guy, who also carries a hawk, escorts him to William's residence at Rouen. On arrival, Harold's hawk is passed to William, suggesting either that it was taken from him or that it had been brought as a gift for William. On his eventual release, Harold returns to England. Now, fast forward to 5 January 1066 when Edward the Confessor dies, childless, and Harold, the king's brother-in-law, is crowned king of England the next day. News travels swiftly back to William in

Normandy, who, as Edward's cousin, assumes himself to be the rightful heir and, after amassing an army, invades England, killing Harold at Hastings on 14 October.

The first person to analyse the birds in the Bayeux Tapestry was zoologist and ornithologist William Brunsdon Yapp in the 1980s. Known as 'Brunny' to his friends, Brunsdon was Yapp's mother's maiden name which he added to make himself 'distinctive' on going up to Cambridge as an undergraduate. After a career as a lecturer at the University of Birmingham, where he was considered a curmudgeon, Yapp's study of birds in medieval iconography was his retirement project. He wasn't the first academic to be interested in medieval birds but his extensive research made a substantial contribution to understanding our relationships with birds in the Middle Ages. He spread his net wide and, as well as scrutinizing the birds in the Bayeux Tapestry, he analysed, identified and reported on birds in misericords, missals, psalters and bibles as well as in the great treatise on falconry written in the early thirteenth century by the Holy Roman Emperor Frederick II.[2]

Over the years, Yapp and other scholars have gently teased apart the threads of the Bayeux Tapestry in search of hidden meanings. Yapp paid particular attention to the species of hawks depicted, in the hope that this might reveal more of what was actually happening. Identification was important because it was thought that the species of raptor one owned reflected one's social rank. Harold's hawk was initially assumed to be a Eurasian Sparrowhawk – among the smallest of the birds employed in falconry – and used here to signal Harold's insignificance. But as Yapp points out, it is clear if one looks at the tapestry that both his and Guy's hawks are far larger than a sparrowhawk, and therefore much more likely to be Northern Goshawks. And, although the idea of a socially enforced link between hawks and status – a kestrel for a knave etc. – has no basis in fact, large, supremely powerful raptors like goshawks and Gyrfalcons automatically conferred greater prestige than a diminutive sparrowhawk or Merlin.[3]

Harold and Guy both hold their hawks in their left hand, and their horses' reins in the right, as was the tradition. This is thought to be the basis of the English riding (and later driving) on the left-hand side of the road. There is also a puzzle: Harold and Guy each hold their hawk on a bare, ungloved hand, something that no falconer now would contemplate.[4]

Completed in the 1070s, the Bayeux Tapestry tells the story of the events leading up to and including the Norman invasion of England. Meticulously observed and executed, seventy metres long and fifty centimetres wide, it was commissioned by Bishop Odo, William's half-brother, and made – embroidered – by expert needlewomen in southern England. In one sense the tapestry seems to me rather like an elaboration of the El Tajo cave paintings, a similarity reinforced by the 200 tiny birds in the tapestry's margins, above and below the main narrative. As Yapp points out, the sheer number of birds in the Bayeux Tapestry is unusual, for it was not for a further two centuries that birds started to appear in abundance in illuminated manuscripts. He speculates about where the images came from: were they copied, drawn from life, or simply from the artists' imagination? Despite the limitations of working in textiles rather than paint, the distinguishing features of species like the crane or the peafowl render them readily identifiable.[5]

Historians have wondered whether those miniatures along the tapestry's border might offer some additional insight into medieval life, including our relationship with animals. Certainly, there are scenes of farming, of a man directing a slingshot at birds and of a raptor in pursuit of a running hare. In some instances, however, the bird marginalia are there to reinforce the main narrative, as in the case of birds flying in the same direction as the invading Norman forces, or, at the point where Harold is captured by Guy in 1064, the birds in the border have their necks tied in knots – the strangulation of Harold's ambition. Prominent also among the marginal birds on the tapestry are those from the moralizing scenes in Aesop's fables: the crow and the fox, warning of the risks of flattery; the crane and the wolf, telling us not to expect a reward for serving the wicked; and the kite and the frog, reminding us of how the treacherous are destroyed by their own actions.

Falconry's origins lie in the distant 'Orient', some time after the last Ice Age, probably between 2000 BC and 750 BC. Rock art dating from 1300 BC in ancient Anatolia depicts raptors with their attendants, and there is mention of falconry in China even earlier, although not everyone is convinced by this.

There's no evidence that the ancient Greeks ever practised falconry as we now know it, and the 'hawking' mentioned by Aristotle in his *Remarkable Things Heard*, comprised:

> [an] occurrence, which is incredible to those who have not seen it. For boys, coming out of the villages and places round to hunt small birds, take hawks with them, and behave as follows: when they have come to a suitable spot, they call

> the hawks addressing them by name; when they hear the boys' voices, they swoop down on the birds. The birds fly in terror into the bushes, where the boys catch them by knocking them down with sticks. But there is one most remarkable feature in this; when the hawks themselves catch any of the birds, they throw them down to the hunters, and the boys after giving a portion of all that is caught to the hawks go home.

Jeremy Mynott has pointed out in his *Birds in the Ancient World* that it would actually make more sense if the beaters were flushing the birds for the hawks to catch, rather than the other way round. Nevertheless, a reciprocal arrangement like this, or one in which men stole prey captured by hawks, may well have been the beginning of hawking.[6]

As one falconry scholar has said:

> Whoever conceived of turning a raptor into a hunting weapon must have seen certain birds catch game faster than he could set an arrow or throw a spear. It could do this at a much greater distance than his weapons reached. Raptors could also spot game far beyond the hunter's eyesight. Whoever imagined that he could harness these awesome talents for his own use must have been something of a visionary.[7]

There is a sliver of evidence that Romans might have engaged with falconry: a fragment of an intriguing mosaic dating from the middle of the Visigothic period in Portugal, around AD 500, of someone holding a hawk – a Eurasian Sparrowhawk or Northern Goshawk.[8] This doesn't necessarily mean the Romans were falconers; the image could well have been that of an exotic visitor whose unusual accoutrement made him worth depicting. After the Romans left

Britain in 410, the invading Saxons brought falconry with them, igniting a passion that would burn undiminished in Britain for over a millennium.

There were two types of Saxon falconers: fowlers, who used trained raptors to catch food for the table, usually for a wealthy employer, and the nobility, mainly Saxon kings, who flew falcons for fun.

In both cases, birds were acquired either as chicks and reared in captivity, or as adult birds taken from the wild and then trained. Training was brutal, comprising control over the birds' food intake, but also 'seeling' their eyelids – sewing them together – so the birds were shielded from any visual disturbance until such time as they were under control. Once a bird was controllable, it was exposed to horses, dogs and people so that it became used to them, and its eyes un-seeled.

There were two classes of raptor: falcons, birds with long, pointed wings that usually attacked their prey from high in the air, killing them through high-speed impact, and hawks, which hunted low over the ground or in woodland with their short wings and killed by grasping. Species like the goshawk have disproportionately large feet, and slow-motion film footage shows how, on making contact with their prey, they and their smaller cousin the sparrowhawk pump their talons rapidly in and out to disable their victim. Goshawks were (and are) typically flown at ducks, pheasants, Rooks and hares. Notwithstanding these two distinct types of predatory bird, the terms 'hawking' and 'falconry' are used interchangeably.

Falconry was at that time the sport of kings, and the nobility's falconers required special attributes. The scholar Adelard of Bath, born around 1080, states that the falconer must be

> sober, patient, chaste, pleasant smelling, free from preoccupations, with drunkenness, the mother of forgetfulness . . . and that visiting prostitutes transmit parasites to the birds when they are touched, a bad breath makes them haters of men and fills them with bad air, so that they suffer from rheum.[9]

The falconer and American academic Tom Cade described falconry as the most 'intellectually demanding and educational form of hunting ever devised', requiring 'a high degree of skill and devotion from the falconer'. He adds that falconry leads to 'a deep appreciation of nature, a practical study of natural history and quite often to serious scientific research on birds of prey'. Cade became head of the Laboratory of Ornithology at Cornell University in the 1960s, when organo-chemicals were decimating raptor populations across both North America and much of Europe. By working out how to breed them in captivity, Cade was responsible for the success of the Peregrine Recovery Program. When he died aged ninety-one in 2019, the writer Helen Macdonald captured the essence of his philosophy as a scientist: 'the long conversations I had with him in the early 2000s . . . showed me again and again that at the heart of his life was that miraculous, beautiful, and astonishingly strong bond he had with raptors'.[10]

Few closer relationships exist than those between owners and their hawks, despite these often being wild-caught birds, trained for hunting and sometimes later released back into the wild. Falconers' birds are rarely 'tame'; rather, they tolerate their owners, and unlike many other birds kept captive such as parrots or corvids, rarely show any affection towards them. Nonetheless, falconers love their birds, and the excitement they generate when hunting was (and still is) the main motivation. As one scholar points out: 'Not only did falconry include fine flights, but it also involved the exhilaration of the kill, the energy expended in retrieving the game, the gusto of the successful venture, and the well-earned post-hunt repose.'[11]

A poignant sense of the bond between a falcon and its owner is provided by the falconer Nick Fox, writing in 1995:

> When a falconer flies his falcon, part of his heart flies with her. This is the difference between the falconer and a man with a gun. The wise falconer can feel the spirit of a falcon, like an aura . . . The spirit of a horse or dog is different [because] they have a desire to please [whereas] the spirit of the falcon is independent, self-reliant and usually stronger than man's. Daily contact with such spirit enhances, inspires and strengthens our own.[12]

Conspicuous ownership of falcons became an important part of upper-class medieval life in Britain, and along with other forms of hunting, falconry was considered a crucial preparation for a military life in what were harsh and aggressive times. Hawks and falcons were status-enhancing ornithological bling, routinely carried around even when there was no thought of hunting – an extension of the owner's phenotype. There are numerous images of amorous young men with a hawk on their fist, single-handedly courting a young woman. It was also said that a knight wouldn't part with his hawk even to procure his own liberty when taken prisoner.[13] Hunting remains a marker of social status today, but more often with guns than falcons, for, ironically, many of those who shoot birds for sport hate and persecute hawks because they interfere with their fun.

In medieval times owning a raptor reinforced rank because their acquisition, training and maintenance were time-consuming and expensive. You sent men out to find eyries and collect nestlings that were then laboriously hand-reared. The alternative was to buy adult birds, trained or untrained, that had been trapped on migration at such places as at the famous Valkenswaard in the Netherlands, or from breeding sites in Iceland or Scandinavia, and imported at considerable expense. Typically in the eleventh century, a Peregrine would cost £1 and a Gyr £2, at a time when a knight's annual income was around £20. The aristocracy employed vassals to rear, train and care for their birds, but even so, it was still part of a young noble's education to be able to fly a hawk. It was not just men; noble women also learned to fly falcons and are often depicted doing so. Such was their value that falcons were regularly given as royal gifts, the apogee being a white Gyrfalcon.

They hardly needed to, but the nobility justified hawking as a healthy diversion from idleness – the result of having too much of everything. More modestly, the Holy Roman Emperor Frederick II suggested that the main benefit of falconry was that it 'enabled nobles and rulers disturbed and worried by the cares of state to find relief in the pleasures of the chase'. Even more astutely he added that 'both rich and poor, by following this avocation [falconry] may earn some of the necessities of life; and both classes will find in bird life attractive manifestations of the process of nature'. Such percipience characterized Frederick, as we will see.[14]

Brunsdon Yapp's interest in medieval birds extended well beyond falconry and those embroidered into the fabric of the famous tapestry. Assiduously, he travelled the length and breadth of Britain, and across Europe, visiting libraries, cathedrals and private collections in search of illuminated manuscripts in which might lie clues for what birds of all types meant to people of the Middle Ages.

One manuscript he examined was a sketchbook dating from around 1400, acquired – with no provenance – by Samuel Pepys sometime in the 1600s. A contemporary catalogue referred to it as 'An ancient book of Monkish drawings and designes for church use', and it subsequently became known as 'the Pepysian sketchbook'. Amidst the pages of angels, apostles and prophets there are eight sheets of bird images, four of them in colour.[15]

The sketchbook is thought to be an extremely rare surviving example of a 'model book' that provided images to be used in other circumstances, including missals, embroidery and stained-glass windows. Some of the species illustrated are readily identifiable, including a white Gyrfalcon. Others are well-known British (and European) birds: the Eurasian Bullfinch, European Goldfinch, Common Crane and Eurasian Jay, but the inclusion of two Continental species, the Ortolan Bunting and Red-legged Partridge (that did not then occur in Britain), suggested to Yapp that the sketchbook was Continental in origin.[16]

Several successive generations of ornithologists had already challenged themselves to attach a name to all the birds in the Pepysian sketchbook, starting with someone unknown who annotated the drawings back in the 1400s. Then there was the foremost Victorian ornithologist, Alfred Newton, whose notes on this are now lost. In 1925 the medievalist scholar and ghost-story writer M. R. James enlisted the help of two professional ornithologists to attach names to the birds, probably on the basis of rather muddy black-and-white photographs rather than the coloured originals. Then, in 1959, the *Illustrated London News* decided to publish the four pages of colour images and invited Britain's top ornithologist, David Lack, to provide identifications.

Yapp, who had looked at the originals, later took issue not only with some of Lack's identifications, but also those in James's earlier account. He sniped at Lack: 'There are several hawks . . . and some rather similar birds which Lack called cuckoos.'

This is slightly unfair, since in two cases, Lack says 'cuckoo (?)' – the question mark clearly indicating his uncertainty, and some of the images are so poor it is impossible to be certain. Together with four other bird enthusiasts, I looked at these images specifically to consider the two birds that James, Lack and Yapp identified as Red-legged Partridges. We all felt that this identification was unlikely because the birds lack several key diagnostic features, but we could offer no alternative. See what you think (see endpapers).[17]

I also categorized the sixty-five birds as raptors, raptor prey, cage birds or 'others'. A quarter are raptors, about half are potential raptor prey (including game birds and the Green Woodpecker), a tenth were possible cage birds (such as the bullfinch, goldfinch and nightingale) and the others comprise a miscellaneous group including the Common Kingfisher, cockerel, peacock, Common Cuckoo and Mute Swan. Overall, two-thirds are birds associated in some way with falconry.

Images of birds were popular long before the Pepysian sketchbook, of course. They exist in Anglo-Saxon manuscripts from the 690s, where the birds are usually depicted in schematic rather than realistic form. In the eleventh and twelfth centuries birds adorned the pages of huge church bibles that were later superseded by more conveniently sized versions.[18] For a few decades either side of 1300 there was an explosion in the abundance of personal psalters. These richly illustrated books of psalms were commissioned by wealthy patrons, and often given as wedding gifts, like the so-called 'Alphonso Psalter' that may have been prepared for the marriage of Edward I's son, Alphonso, to Margaret of Holland. The wedding was scheduled to take place in late 1284, but the unfortunate Alphonso died before the happy event and the psalter remained unfinished. The 'Bird Psalter', possibly a companion volume, and probably also created for the lucky Margaret, contains excellent images of at least twenty-three identifiable bird species.[19]

These extraordinarily illustrated artefacts reveal a new appreciation of birds. Perched upon or flying around the margins of the text, they are sometimes merely decorative, while in other cases they form scenes from real life, such as netting partridges, or using a decoy owl to lure small birds onto birdlime sticks. In other instances the birds carry a symbolic meaning, most notably the dove, signifying the Holy Ghost, and the peafowl, whose flesh was (erroneously) said to resist decomposition, signifying the resurrection. The difference between perched versus flying birds is intriguing: images of perched birds coincide with a period when keeping birds in cages was popular in Britain, allowing artists to obtain a decent view of them. Flying birds occur slightly later, when Continental rulers had large aviaries, and the occupants could be seen on the wing.[20]

As the popularity of personal psalters was waning in England in the mid-1300s, a new type of illuminated manuscript – alive with avian imagery – was appearing in

France. Books of Hours were devotional manuals that allowed the laity to mimic monks by identifying devotional services appropriate for different times of day.

One of these, Catherine of Cleves's Book of Hours, commissioned for her marriage – aged just thirteen – to Arnold, Duke of Guelders in the Netherlands in 1430, provides clear evidence for the nobility's enthusiasm for caged birds. One page shows a variety of cage types, a string-pulling goldfinch and an evil-looking cylindrical cage turned by a man. Rather than being merely decorative, as the birds are assumed to be, this cage may be symbolic, representing the womb and hence Catherine's reproductive responsibilities as a wife and a woman.[21]

Two major religious themes feature prominently in Books of Hours: the Creation and the Apocalypse. The Creation is the beginning of the earth, as described in the Bible in Genesis, in which Adam names all the animals created by God. Medieval depictions of this event usually show Adam with a variety of mammals (such as big cats, deer and domestic animals), and lots of birds, with reptiles, fish and other taxa represented much less often. The abundance of birds presents a paradox since the identification and naming of different species in the medieval period was a source of considerable confusion. Presumably, Adam knew what he was dealing with, but the identity of some birds remained as one of God's great riddles that became an enduring challenge for ornithologists over the subsequent centuries.

The Apocalypse, also known as the Revelation, is generally interpreted as the end of the world – followed by the Day of Judgement: 'And I saw an angel standing in the sun; and he cried with a loud voice, saying to all the fowls that fly in the midst of heaven, come and gather yourselves together . . . that ye may eat the flesh of kings, and . . . of horses, and . . . of all men.'[22]

As with so much in the Bible, this passage has been subject to a wide range of interpretations, and the various medieval images of the Apocalypse may also have different meanings. One is that it reflects a broad view of history. Another is that it refers to specific historical events, like the fall of the Roman Empire, thought to be a consequence of the Romans' depravity, epitomized by their slaughter of both animals and people for public entertainment in their amphitheatres. Another meaning is allegorical, representing the ongoing struggle between good and evil. And perhaps the most telling is the current interpretation: a warning of what is to come as a result of the way we have treated the earth.

These medieval images were designed to haunt the viewer

A medieval Apocalypse (mid-thirteenth century) with flocks of birds feasting on the dead (courtesy Bodleian Library).

and have never been more apposite than they are today, as we are challenged by new diseases, as we anticipate the concomitant global economic recession, and as we watch climate change continue apace and the last of the world's natural resources disappearing into the pockets of greedy developers.

Medieval depictions of birds in the Apocalypse comprise two very different types of image: either St Francis preaching to the birds, or 'The Call of the Birds', as the scene in the Apocalypse is sometimes referred to, followed by human corpses being consumed by scavenging birds.

Many animals feature in apocalyptic paintings, but birds are the most frequent, precisely because they held such a special place in the medieval mind, as they had done for millennia. Hovering midway between heaven and earth, half angels, half animals, the birds were the ultimate messengers

The Queen Mary Apocalypse, or 'Call of the Birds', with an angel, a number of readily identifiable birds and a rabbit in its burrow, an image dating from 1325 (courtesy British Library).

linking the present to the afterlife. The singular status of birds made them especially suitable for the condemnation of humans on the Day of Judgement.

How was St Francis of Assisi drawn into these apocalyptic scenarios? By sleight of hand, it seems. He is said to have preached to the birds in the Spoleto valley, telling them of God's love for them. This story may be true or may simply be allegorical. Either way, the saint's well-known respect for the natural world and his unconventional view that all living creatures, and not just people, were part of God's plan, makes him a powerful symbol, and the reason why his followers (the Franciscans) made him equivalent to the archangel. This is how he appears in many images, and why, in 1979, he was designated the patron saint of ecology by Pope John Paul II. In apocalyptic images of St Francis preaching to the birds, he is effectively asking, 'What have they done to the earth?' What have *we* done to the earth?

Throughout his career Yapp seems to have pursued an eclectic range of interests, writing about Britain's biological productivity, ageing, long-distance walking, the National Parks and driving his Bentley around the Lake District. Indeed, from an ornithological perspective, it may well have been the fact that he specialized neither in a particular bird species nor a particular topic that made him seem a jack-of-all-trades and prevented him being considered part of the ornithological elite. We do not know what first drew Yapp to medieval manuscripts, but once his interest was piqued, it was inevitable he would eventually find himself looking at the birds in one of the most remarkable of all medieval documents, *De arte venandi cum avibus* – *On the Art of Hunting with Birds* – written by Frederick II.

Frederick was ostensibly the child of Henry VI and Constance, queen regnant of Sicily, but there was some doubt about his true parentage. Constance was forty when he was born and it was rumoured that she had faked her pregnancy, with the new-born Frederick smuggled into the royal palace. Like his father, Frederick duly became king of Germany, Holy Roman Emperor and king of Sicily. He proved exceptional in many ways and was dubbed the *stupor mundi* – the astonishment of the world. Frederick was unconstrained by any religious beliefs and Pope Gregory IX considered him to be the predecessor of the Antichrist, so he excommunicated him. Intellectually curious, perceptive and unorthodox, Frederick was a keen falconer, but he was unique in his day in recognizing that, to understand birds of prey, he had to understand birds in general. His extraordinary manuscript covers the anatomy, ecology and migration of birds, fully justifying the later claim that he was the father of ornithology.

The Art of Hunting with Birds was produced between 1244 and 1250. Despite being known by repute, the book was curiously out of reach to those interested in birds since it remained unpublished until 1596, and even then was probably not easily accessible. The English version was translated from two manuscripts, the oldest one – known as the Bologna manuscript – probably created during Frederick's lifetime, and the Vatican version, which contains an extraordinarily beautiful series of iconic illustrations.

Casey Wood, an American ophthalmologist, undertook this English translation during the 1930s and 40s. If the link between ophthalmology and medieval manuscripts seems unlikely, let me explain. Wood's career was dedicated to curing human eye conditions. Well aware that the eyesight of birds, notably raptors, was far superior to our own, he developed an interest in avian vision that in turn lured him towards falconry. In fact, Wood's interests were wide-ranging, including concern for the health of his busy fellow medical practitioners, whom he urged – for relaxation – to read, for this, he said, will 'allow you to come in contact with the most distinguished people in all humanity'. Wood read – a lot. In the 1930s he produced a bibliography of zoology, an annotated list of all the zoological works in a library he had established at McGill University in Montreal, some 20,000 items in total. Well aware of Frederick's manuscripts, Wood had purchased several of the published versions. It was on a visit to Rome in 1930, however, that he was able to see the Vatican manuscript, with its 600 exquisite miniature paintings of birds.[23]

Casey Wood's translation was produced with the help of his niece, Florence Marjorie Fyfe, who transcribed and edited the text for the publisher; but it was Wood who undertook the mammoth task of converting the Latin into English. Wood and Fyfe's imposing 'brick' of 617 pages was published in 1943 – fortuitously, just as falconry was lifting off again as an elite hobby in North America after years of 'neglect'. The book received rave reviews.[24]

Most people who see Frederick's manuscript marvel at those illustrations. Unique in their style, they may have been based on sketches by Frederick himself, and there's something strangely beguiling about their naivety. They have been enthusiastically described as 'amazingly exact', 'extraordinarily life-like' and 'the first truly realistic representation of birds'.[25] But none of those who heaped praise on these tiny images was either an ornithologist or as familiar with medieval bird images as Yapp, and he – without wanting to take

Illumination from a French copy of Frederick II's manuscript *De arte venandi cum avibus* showing falconers spraying water from their mouths onto their birds as a way of calming them, and making them feel as though they have just bathed.

anything from Frederick's ornithological achievements – disagreed. Only 10 per cent, he says, of the seventy or so species depicted, are drawn accurately enough for unambiguous identification. The birds of prey are the most ambiguous. The reason, he presumes, is because the final images were created after Frederick's death, and by at least three different artists, none of whom can have been especially familiar with birds. Notwithstanding their lack of ornithological precision, the images that accompany Frederick's text are what help to make his manuscript so seductively attractive.[26]

To understand the significance of Frederick's ornithology – built around a combination of his secular approach and his knowledge of Aristotle's works – we must go back to the beginning of the first millennium to find the early course of natural history.

With the coming of the Christian faith, the pursuit of secular knowledge was considered ungodly. In the Middle East, what had once been a thriving interest in the works of Aristotle was abandoned and replaced by mysticism and symbolism drawn from what was then 'the Orient' and from Greece itself. This created a culture in which anything that did not coincide with religious dogma was assigned a symbolic role. It was on this basis that a group of Christian scholars around AD 200 produced the *Physiologus* – an animal encyclopedia consisting of lessons from the Bible presented in a way that ordinary people could easily understand. Birds were bestowed with magical or moral properties: the phoenix rising from the flames representing the resurrection, or the pelican piercing its own breast to feed its young with its blood in an act of self-sacrifice. These stories, accepted as true, conditioned people to the supernatural, predisposing them to accept unquestioningly many of the doctrines of the scriptures.[27] An early example of fake news in the service of religion.

Despite the Christian dominance, there were pockets of secular resistance. Anicia Juliana, daughter of the Roman emperor Anicius Olybrius, for example, in the sixth century commissioned a manuscript with over 400 images of animals and plants. The text is derived in part from the work of the first-century physician Dioscorides and partly from the writings of Dionysius [Periegetes?], and, regarding birds, focuses mainly on methods of capture. This is the 'Vienna

Dioscorides', which Yapp examined, and includes some striking colour images, including a Black Francolin, Golden Oriole, Blue Rock Thrush, Eurasian Coot, Ruddy Shelduck, and Common Ostrich – all remarkable for their accuracy.

In the seventh century St Isidore, Archbishop of Seville, said to be the last scholar of the ancient world, created an encyclopedia of universal knowledge, *Etymologies*, designed for instructing students and which included a volume on birds. Despite what we assume were Isidore's best intentions, his ornithology demonstrates just how corrupted secular knowledge had become by this time. The swan, he says, sings sweetly because of its long curving neck; the cuckoo migrates on the shoulders of kites, and the saliva of cuckoos creates grasshoppers. On the other hand, he does state correctly that the cuckoo lays its eggs in the nests of other birds, information obtained – untainted – from Aristotle's writings.

It was Spain that became the source of Aristotle's revival. In the country's southern regions, invaded by North African Moslems in the eighth century, Arabic culture subsequently flourished. In the twelfth-century the Andalusian polymath Averroes, whose novel philosophy that man was made to discover the truth, triggered a renewed interest in the natural world. Averroes had obtained Syriac translations of Aristotle's works and in promoting his ideas attracted the attention of Scottish scholar Michael Scot, who was later employed by Frederick II to translate Aristotle's works into Latin.

One of those to benefit from Michael Scot's translation was a contemporary of Frederick's, the German Catholic bishop Albertus Magnus, arguably the greatest philosopher of the Middle Ages. His book *De animalibus* is an account of Aristotle's studies of animals, annotated with Albertus's own comments and observations. Such was Albertus Magnus's knowledge of birds that in 2014 the ornithological historian Jürgen Haffer declared him one of the fathers of ornithology.[28]

Because falconry was expensive, time-consuming and, above all, useless – the ultimate signal of conspicuous consumption – it is not surprising that it eventually turned round and lunged at its practitioners. Medieval nobles in Europe, like much of the prey they pursued with their Gyrs and Peregrines, were soft targets.[29]

The lower classes, excluded very obviously from participating – other than as lackeys – despised the falcon-fisted nobility for their right to ride their horses and dogs through their crops in pursuit of their hunting birds. Medieval falconers were often arrogant, as John of Salisbury, writing in 1159, complains: 'Rarely is [a hunter] found to be modest or dignified, rarely self-controlled, and in my opinion never temperate.'[30]

Some of the greatest derision for falconers appeared in the moralizing bibles produced in Paris in the 1220s and 1400s. Each page comprised eight tiny images arranged in two columns of four; pairs of biblical and moralizing text and pictures, in which the falcon is a symbol of 'sin, worldliness or improper clerical behaviour'. Falcons are portrayed alongside money-changers, images of gluttony and lust in scenes depicting 'idolatry, inconstancy, contumacy, greed and pride'. Elsewhere, pride is described as riding a lion and carrying an eagle; envy carries a sparrowhawk and avarice a hawk.[31] Hieronymus Bosch, in his painting on a wooden table top of the seven deadly sins, depicts Envy as a merchant looking longingly at a man with a falcon. Similar raptor-related moralizing images also occur in stained-glass windows and in tapestries.

The Middle Ages were particularly heartless times for people, but also for hawks and especially for their prey. Despite falconers being urged to love their birds and use 'nothing but gentlenesse' in their instruction, the training, in some hands at least, was a cruel business, and involved breaking the bird by depriving it of sleep. This was the method T. H. White found in Edmund Bert's *Treatise of Hawks and Hawking* of 1619, and assumed it to be the best way to train his Northern Goshawk in the 1930s.

White writes:

> In teaching a hawk it was useless to bludgeon the creature into submission . . . the old hawk masters had invented a means of taming them which offered no visible cruelty, and whose secret cruelty had to be borne by the trainer as well as by the bird. They kept the bird awake . . . for a space of two, three, or as much as nine nights . . . All the time he [the trainer] treated his captive with more than every courtesy, more than every kindness and consideration.

After several mutually sleepless nights White enters his mews, where the goshawk 'regarded me with tolerant contempt. He had no doubts about who was the slave, the ridiculous and subservient one who stood and waited.' When the bird finally deigns to jump onto White's hand and eat, it is 'an exultation! What a bursting heart of gratitude and triumph . . . the rest of the day was a glow of pleasure.' But the battle between man and bird was far from over.

I suppose White's goshawk was lucky he hadn't 'seeled' its eyelids as part of the breaking process, but mercifully that unpleasant practice, even in 1619, had long since been replaced by the hood – an innovation from the Arab world. But it did so only because it was more convenient and not to save the hawk any discomfort.[32]

Across Britain and central Europe the nobility typically employed falcons to capture Common Cranes and Grey Herons, birds far larger than either the Peregrine, or the bigger Gyrfalcon, would normally ever take, and they had to persuade them to do so. The training process was pitiless, with the falcon first allowed to attack a deliberately disabled walking crane with its beak tied. From there the raptor graduated slowly to a stronger walking crane, then to a running crane, then a blindfolded flying crane, until finally it was confident enough to catch a free-flying bird.[33]

It was the spectacle of a falcon diving (stooping) swiftly onto the ungainly, broad-winged, long-necked crane that thrilled the hunting parties. Once attacked, a crane simply tried to out-fly its pursuer, occasionally lashing out with its feet.[34] In contrast, herons usually turned to face the falcon as it approached and tried to drive it off with its beak, twisting and turning in the air while also struggling to make an escape. The fact that a crane or heron could fight back against the falcon added to the falconers' excitement.

Cranes, which were routinely served at medieval feasts, were also obtained by other, equally diabolical means. Birdcatchers made paper cones lined with birdlime (glue), with a piece of bait inside their tip, that were then placed inside tiny pits in the fields. Tempted by the bait, the crane placed its head inside the cone – which immediately stuck fast – 'hoodwinking' itself. In a desperate attempt to escape, the hapless crane took to the skies, flying vertically upwards until eventually, exhausted, it fell back to earth, where it was easily retrieved.[35]

As for heron hawking, this is what John Shaw, writing in 1635 said:

> The heron or hernshaw . . . hath a marvellous hatred of the hawk, which hatred is duly returned. When they fight above in the air, they labour both especially for this one thing – that one may ascend and be above the other. Now, if the hawk getteth the upper place, he overthroweth and vanquisheth the heron with a marvellous earnest flight.[36]

What this means is that the falcon (hawks were rarely used to hunt herons or cranes) needed to gain greater height than the heron so that it could stoop onto it at sufficiently high speed to disable or kill it. For the falcon (and falconer) the real danger was on the ground if the heron was not killed outright, for a wounded heron would strike out with its dagger-like bill at lightning speed towards the eyes.

> Whoo-whoo-o-p! Down they come. Down they all three [two falcons and the heron] go together, till, just before reaching the ground, the two old hawks [falcons] let go of their prey, which falls bump. Before he has had time to recover himself . . . the hawks are on him . . . [one] on the neck and [one] on his body. Hurrah for the gallant hawks![37]

The falconer arrives in time to save the heron's life; gives the falcons a pigeon to eat; and 'puts the heron between his knees in a position that he can neither spike him nor the hawks with his bill'. The heron's two long, black head plumes are taken as a trophy, 'a badge of honour' – a ring is put round its leg to identify its captor, and it is then let loose, to fly another day.[38]

A noble sport?

Yet, surprisingly, one of the classic hunting texts from

1575, *The Noble Art of Venerie or Hunting*, by George Gascoigne, contains several poems expressing concern about cruelty. Not to birds, it is true, but for the otter, fox, hart and the hare. He asks: 'Are minds of men, become so void of sense that they can joy to hurt a harmless thing?' Strange indeed that this should have been included in what is otherwise a hunting handbook. Gascoigne may have done so to challenge the deeply ingrained idea of hunting (and hawking) as a celebration of manliness and bloodlust, in what historian Catherine Bates has said articulates 'the classic humanist paradox whereby the rabid huntsman appears more savage and bestial than the gentle beast he pursues'. I have not found anything equivalent expressing sympathy for the victims of hawking, other than the curious symbolism of the despairing male lover as the falcon's victim.[39]

Concern over animal cruelty had been expressed by classical moralists such as Plutarch and Porphyry, but was all but ignored. In the twelfth century the philosopher John of Salisbury asserted that hunting had a brutalizing effect on the character. Indeed, rather than being concerned by the suffering experienced by the Peregrine's prey or the hounds' hare, the emphasis was on how hawking and hunting negatively affected the falconer himself. The human perspective became the dominant view among the few who thought about animal cruelty. Overall, as Dix Harwood says in his book *Love for Animals and How It Developed in Great Britain*, 'the evidence for the sympathetic interest in animals before 1700 is very slight'.[40]

What changed? The subsequent shift to a more compassionate position was thought to be the result of a 'gradual elevation of the moral standard . . . that expanded the area in which human feelings were allowed to operate' rather than any increase in either knowledge or 'definitive reasoning'.[41] It started, as we've seen, with concern over what cruelty to animals might do to the perpetrators. One of the most significant contributions to the debate was *Dives and Pauper*, a fifteenth-century commentary on the Ten Commandments, in which a rich man (Dives) requests instructions regarding the Ten Commandments from his poor interlocutor. The sixth commandment, thou shalt not kill, excludes killing animals for food or those that are 'noxious to man', but does prohibit the killing of animals for cruelty's sake or out of vanity. 'Men should have ruth [pity] of beasts and birds, and not harm them without cause' . . . and, as Keith Thomas has said: 'this is a notable passage and a very embarrassing one to anybody trying to trace some development in English thinking about animal cruelty. For here at the very beginning of the fifteenth century we have a clear statement of a position that differs in no respect whatsoever from that of most eighteenth-century writers on the subject.'

Slowly, slowly, things began to change. Henry VIII's chancellor, Sir Thomas More, considered hunting the 'lowest, the vilest, and most abject part of butchery', and in 1603 a preacher told his flock that there would be no hawking in heaven. From the mid-1700s, the trickle of anti-cruelty literature became a fast-flowing stream – fed further by the work of poets and nineteenth-century writers like John Clare – such that by the late 1800s concern for animals had become a distinctive part of English middle-class culture. Nonetheless, 'Opinions on cruelty by the all-decisive gentry took longer to change.'[42]

Falconry reached its peak in popularity sometime in the 1500s, but once the musket (or 'fowling piece') was invented and found to be so much more efficient, falconry went into decline. In the mid-1600s Charles II was said to have done his arduous best to revive it, because, he felt (and possibly with an ulterior motive), that falconry: 'is most convenient for the ladies'.[43] Despite Charles's efforts, falconry continued to fade, and Giles Jacob in his *The Compleat Sportsman* in 1718, commented: 'The diversion of hawking, by reason of the trouble and expence in keeping and breeding the hawk . . . is in great measure disused: especially since sportsmen are arrived to such a perfection in shooting.'[44]

A few individuals continued hawking during the 1800s and 1900s, but it was the 1970s that saw a real resurgence in hunting with birds of prey in Britain, parts of Europe, the Middle East and North America. In the 1950s and 60s the diminutive American Kestrel had been successfully bred in captivity, enabling researchers to better understand the disastrous effects of DDT and other toxic chemicals on their reproduction. This, in turn, led to a boom in captive raptor rearing, providing Peregrines, gyrs and goshawks for falconers and also, in the Arab world, for falcon racing. Accompanying this, considerable thought has been given to the ethics of rearing, training and maintaining raptors. The standards among 'professional' falconers – those for whom falconry is a way of life – are now extremely high. As for the issue of cruelty experienced by the prey, the main arguments are that falconers are simply allowing their birds to do what their wild counterparts would do; the success of hunts is low because prey have every opportunity to escape; and in contrast to shooting, prey rarely sustain sub-lethal injuries and die a lingering death because hawk hunts are either successful or not. Falconers also emphasize the fact that their birds, when flying, are free to leave whenever they wish.[45]

Medieval falconry bred a careless indifference to cruelty – a position reinforced by the Christian Church's belief that man had dominion over animals and that those animals were incapable of sharing the same physical sensations as ourselves. The Christian attitude to non-human animal life contrasts sharply with that of some ancient Indian faiths, such as Buddhism and Jainism, in which animals are assumed to be sentient and treated with respect. Yet not everyone in the Christian world shared the same views. The Middle Ages saw a tiny number of individuals kick back against falconers, concerned not only about cruelty per se, but also about the way it could corrupt, creating the faintest suggestion in the clouds on the distant horizon that cruelty to animals might be morally wrong. In the following chapters we will see how this particular horizon starts to brighten, creating a fundamental shift in our relationships with birds. Compassion and morality – in certain societies at least – changed very slowly. But here was a beginning.

I saw Brunsdon Yapp only once, but he left a lasting impression. I was at a conference in the 1970s when he had a very public altercation with an eminent ornithologist in which Yapp denied the catastrophic effects of pesticides on Peregrines. To my dismay, it was later revealed that Yapp was on the payroll of the pesticide manufacturers. A contrarian with a reputation for grumpiness, Yapp died in obscurity in Cambridge on 12 March 1990, aged eighty-one. There were no obituaries. The root cause of his unhappiness and cantankerous manner was probably the fact that in 1945, within five years of his marriage, his wife Bridget died, aged thirty-seven. From the moment his wife became ill, Yapp farmed out his two-year-old daughter and lived 'with a series of housekeepers, with whom he quarrelled', finding solace, perhaps, in his Bentley, fine wine and medieval manuscripts.[46]

It may be no coincidence that Yapp's ornithological interests settled on an era once referred to as the Dark Ages. Notwithstanding his obvious contribution to our understanding of the ornithology of this period, how much better it would have been for him had he also looked ahead – as we will do now – to the Renaissance.

5. Renaissance Thinking: The Parts of Birds

Shall I marry, or devote my life to study?
Alessandra Scala (1491)[1]

Amidst the mass of unpublished notes and drawings left by Leonardo da Vinci is a throwaway comment on one of his to-do lists, that says: 'Describe the woodpecker's tongue'.

Other items on Leonardo's to-do lists include discovering why the sky is blue, how the webbed foot of a goose works and how birds fly, but the woodpecker's tongue is the most intriguing.

It certainly captivated Walter Isaacson, Leonardo's recent biographer: 'When I first saw his entry about the woodpecker, I regarded it as most scholars have, as an entertaining oddity – evidence of the eccentric nature of Leonardo's endless curiosity ... But ... Leonardo knew that there was something to be learned from it.'[2]

Although Leonardo left no anatomical drawing of the woodpecker's tongue, and there is no further mention of it in his notes, he must have seen either an image of a woodpecker, or a dead one, with its elongate tongue protruding from its beak.

There are several species he might have come across; the Green, Lesser Spotted and Great Spotted Woodpecker, as well as the Eurasian Wryneck (which is also a woodpecker), all occur in Italy, and they all have a long tongue. The Green

Woodpecker's and wryneck's are the most spectacular, since their tongues extend three or four times the length of the bird's beak.

I have been unable to find any drawing of a woodpecker's tongue that Leonardo might have seen. This may not be too surprising, since prior to the Renaissance there were few images of birds that depicted anything other than their external appearance. Leonardo may have seen a woodpecker captured by a falconer's Peregrine, or one on sale from a poulterer's stall. A painting once attributed (incorrectly) to Caravaggio of a market stall laden with dead of birds, dating from the mid-1500s, includes four woodpeckers: Green, Lesser Spotted, Great Spotted and a Eurasian Wryneck, and we know from other sources that woodpeckers were eaten, at least occasionally.[3]

Leonardo was intrigued by the woodpecker's tongue because he was obsessed with how things worked: he was an engineer, after all, and, as Walter Isaacson says, he was 'over-endowed with . . . curiosity and acuity'. I think Isaacson's bewilderment at Leonardo's fascination for this extraordinary avian structure reflects the chasm that still exists between the sciences and the humanities. A scientist would not have reacted in the same way to Leonardo's comment: they would have accepted that there are some extraordinary and fascinating animal anatomies. In contrast, if you are not a scientist (or an ornithologist), then the idea that anyone would be interested in a bird's tongue would seem bizarre. Isaacson has no problem with Leonardo's interest in the flight of birds, because he knows that we would all like to fly, but tongues?

Then, finally, he understands. 'After pushing myself to be more like Leonardo,' Isaacson says, 'I realized I had become fascinated by the muscles of the tongue. All of the other

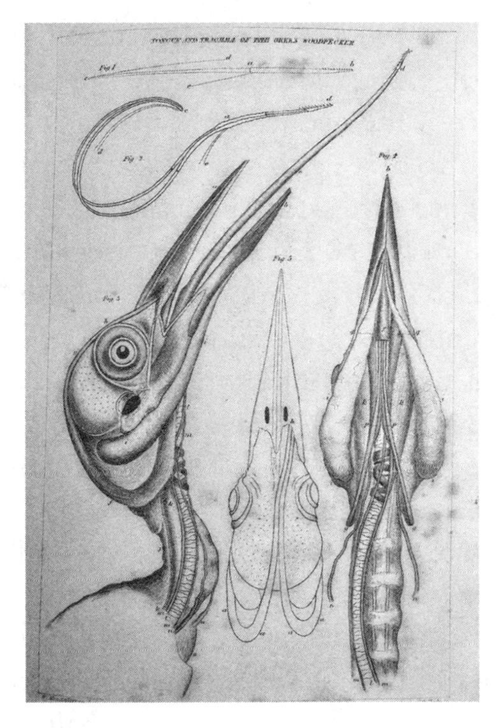

The extraordinarily long tongue of a Green Woodpecker (from MacGillivray, 1852).

muscles he studied acted by pulling rather than pushing a body part, but the tongue seemed to be an exception.' This is true of the tongues of humans and other mammals, but the sight of a woodpecker pushing out its tongue is truly extraordinary.

Once, as part of a public lecture on Leonardo, I dissected a woodpecker to demonstrate its wonderful tongue. My specimen was a Green Woodpecker that had been found dead in a park many years previously and lain inert inside a museum freezer ever since. Using a fine-pointed scalpel, I removed the skin from its head to reveal, first, the bird's huge salivary glands whose secretions lubricate the tongue and provide the 'glue' for the tongue's tip, with which the bird secures its insect-larvae prey. Folding back the skin from the top of the skull exposes the two extremely elongate bones of the tongue – the hyoid bones – whose ends disappear into the right nostril. These two skinny bones run side by side inside a tough tube from the nostril and back over the skull, where they divide, each one then running round the underside of the skull and into the throat and mouth. I opened the bird's beak, grasped the tip of the tongue and gently pulled: the audience gasped at the sheer length of tongue that emerged. I was as thrilled as they were. Just to put this into perspective, the tongues of most other birds are tiny and rooted firmly inside the mouth, not wrapped around the skull.

So extreme is the woodpecker's tongue that creationists assert that it could not have evolved – as they say – by 'chance', nor in small increments, and therefore must have been created by God. What use would half a tongue have been, they ask, ignorant of the fact that there are woodpecker species with tongues of different lengths all over the world.[4]

Leonardo stood astride the dogmatic scholastic era of the past, where the written word – of Aristotle and Pliny, for example – was the ultimate authority, and the more liberated Renaissance, characterized by novel technologies, new ways of thinking and observation of natural phenomena. The invention of moveable type and the production of printed books revolutionized and democratized knowledge. The rediscovery of the Greek philosophers had a similar effect on art, architecture and science, with a greater emphasis on observation rather than dogma, and a more relaxed attitude to dissection. The discovery and exploration of foreign lands broadened horizons and accelerated the accumulation of new knowledge. In Europe, an increasing human population required more food and more efficient agriculture.

All of these changes altered our relationships with birds. The dissection of people and other vertebrates, including birds, and invertebrates encouraged an interest in the relationships between different types of organism. This comparative approach is beautifully captured by Pierre Belon's illustration from 1555 showing the similarities between the skeletons of a bird and a human.[5]

The Greeks had dissected and described the internal anatomy of the human body and Aristotle examined the internal

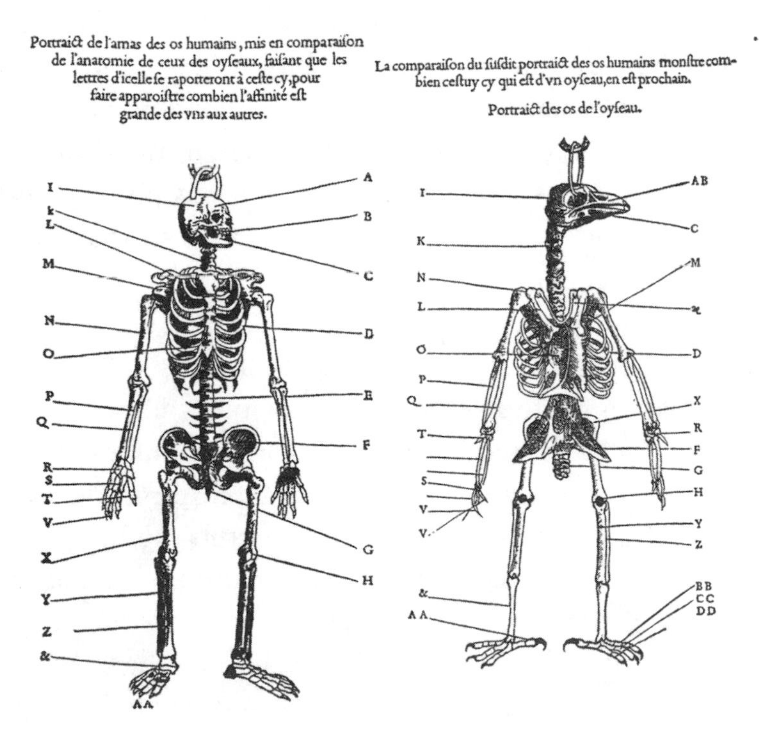

Pierre Belon's comparison of human and bird skeletons (from Belon, 1555).

anatomy of nine different species of birds.[6] He distinguished between the crop (a storage structure) and the gizzard (the true stomach), but incorrectly assumed both to be for storing food, with digestion taking place elsewhere. Aristotle may also have identified the air sacs in birds – which we now know form part of their super-efficient respiratory system. By the second century AD, human dissection had gone out of favour, and with the emergence of Christianity in Europe 'the development of rational thought and investigation was paralysed' and human dissection prohibited. However, in the thirteenth century, Frederick II issued a decree that 'a human body should be dissected at least once in every five years for anatomical studies, and attendance was made compulsory for everyone who was to practise medicine or surgery'.[7] This eventually resulted in the first legal human dissection, in the University of Bologna in 1315. These early demonstrations, however, were less enlightening than one might imagine. The dissection was directed by a professor sitting apart on an elevated dais, reading aloud from an anatomical text. The actual dissection was performed by a barber-surgeon, with an 'ostensor' ('one who shows') pointing out the relevant bits to the audience. The limitation was that everyone saw whatever was read from the text, regardless of what was apparent in the dissection itself.[8]

Relaxation of the restrictions previously imposed by the Church allowed a flowering of anatomical studies. Florence was the epicentre, with Leonardo at the forefront, dissecting the muscles that birds employ for flight in the hope that he might find a way of launching humans into the air. Except by word of mouth, Leonardo's anatomical efforts had little impact because he never published his findings. As Walter Isaacson says, Leonardo's 'passion for acquiring knowledge was not matched by one for sharing it widely'. Nonetheless, others eagerly adopted this new, more intimate way of looking at life (or death). As Belon had anticipated, these anatomical studies showed just how similar we are to birds.[9]

To prevent people making too much of that similarity, the brilliant and influential Renaissance philosopher René Descartes used his persuasive writings to reinforce the medieval mantra that animals and humans were distinct. Animals, he said, lacked a soul, and without one they were incapable of feeling pain or emotions such as fear or anxiety. To Descartes, animals were mere automata, and this was a view that the Church was happy to continue promoting. The abuse and misuse of animals therefore persisted throughout the Renaissance and beyond, with vivisection and all its desperate cruelties part of the discovery process.

To us this seems like a strange contradiction: humans and non-humans were anatomically similar yet spiritually distinct. But to those of the Renaissance there was no contradiction, it was merely part of God's great plan, a plan he posed as a puzzle for people to decode. There's an irony here too. From the time of his death around AD 200 until the Renaissance, the Roman physician Galen had been the unassailable authority on human anatomy. Yet in 1541, close reading of his text by the Flemish physician Andreas Vesalius revealed that Galen's work was based entirely on non-humans. He had dissected Barbary apes rather than people, a substitution necessary because human dissection was forbidden in Rome. He had assumed we were identical. We are similar, but not identical, and as a result, Galen's works contain some errors.

Vesalius was Professor of Surgery and Anatomy at Padua, and as part of his teaching performed human dissections himself rather than delegating to a barber-surgeon assistant. This first-hand experience allowed him to see Galen's mistakes, and even though – or perhaps because – he was engaged in preparing a new edition of Galen's works, he decided to produce his own *De humani corporis fabrica libri septem* (*On the Fabric of the Human Body in Seven Books*) in 1543. It was Vesalius's magnificently clear and original woodcuts – the kind of innovative visual representations that characterized the Renaissance – that sold his books, making him famous throughout Europe. Vesalius rectified many of Galen's mistakes, but left

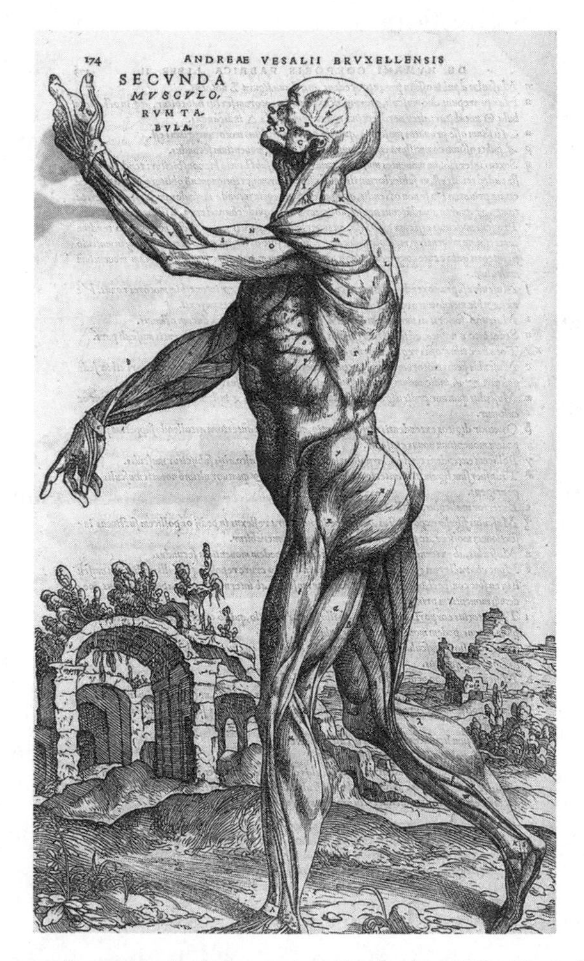

One of the superb anatomical illustrations in Andreas Vesalius's *De humani corporis fabrica* (*On the Fabric of the Human Body*) from 1543.

a lot uncorrected, and despite its undoubted brilliance, the text is largely based on what Galen had written. As one commentator said, it is strange that Vesalius 'could have disclosed so much and not discovered more'. Part of the explanation is that Vesalius was not inclined to draw comparisons between humans and other animals: he was not a comparative anatomist.[10]

No such inhibitions plagued Volcher Coiter, a Frisian scholar who also worked at Padua, some thirty years after Vesalius. He trained under some of the great names of Renaissance medicine: Gabriele Falloppio, Bartolomeo Eustachi, Ulisse Aldrovandi and Guillaume Rondelet. Directed by Aldrovandi – who later produced one of the great ornithological works of the Renaissance – Coiter dissected a huge range of animals, from tortoises, crocodiles, lizards and snakes, through pigs, goats, horses, wolves, foxes, hedgehogs and bats to an interesting array of birds, including a Great Northern Diver, Great Cormorant, Common Crane, chicken, Common Starling, Eurasian Wryneck and Great Spotted Woodpecker.

At the age of thirty, in 1564, Coiter was appointed Professor of Surgery at Bologna. His life there was far from easy. By openly criticizing the inappropriate activities of barber-surgeons he made enemies and, jealous of his success, even his colleagues plotted against him. His conversion to Protestantism made him especially vulnerable, and 'after a street fight by students [Coiter] was taken prisoner, bound and sent to Rome'. A year later, someone – we don't know who – helped secure his release. In 1569 he was appointed town physician in Nuremberg and despite being 'a broken man, unhappy and humbled by his cruel experience', it was here that he published important medical works and his groundbreaking anatomical studies of birds. Indeed, Coiter was the man who filled the void left by Leonardo by providing the first illustrations – in

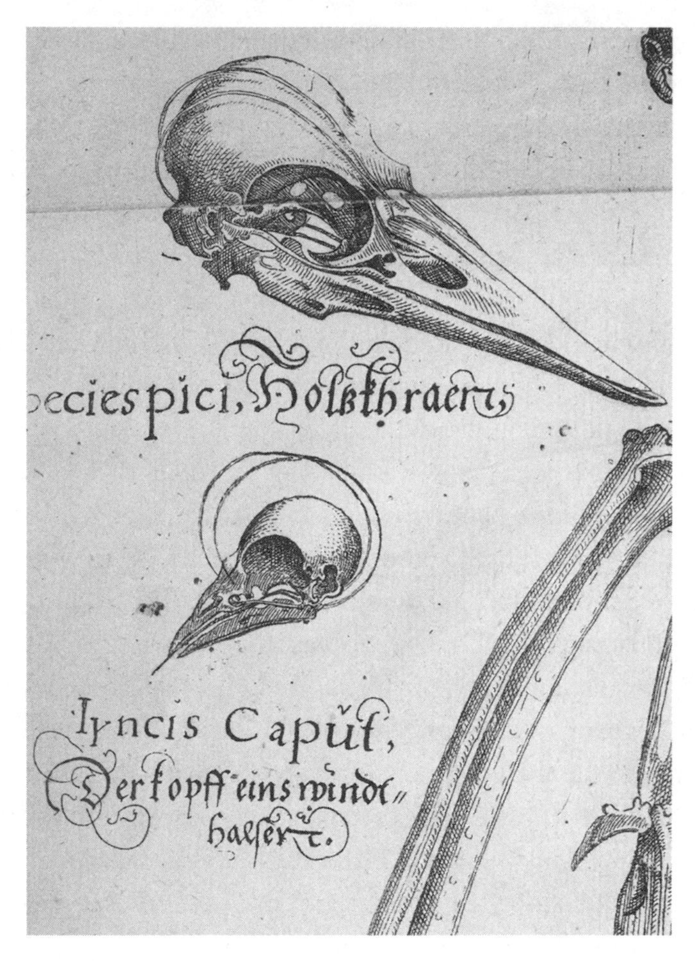

Volcher Coiter's woodpecker and wryneck tongues with their extraordinarily long hyoid bones running back through the mouth, behind the skull and attached in one nostril.

1575 – of the woodpecker's long tongue, as well as an account of the muscles that extend and retract it.[11]

While writing his bird encyclopedia in the 1550s, Pierre Belon, whom we just met, obtained the dried head of a Toco

Toucan. He correctly noted the lightness of the skull, and that the jagged-edged beak seemed to lack nostrils. With nothing else to go on, Belon had to imagine what the rest of the bird looked like if he was to fit it into his classification scheme. The only other birds he knew of with saw-like bill edges were mergansers (the eponymous sawbill ducks) and so – reluctantly, one hopes – Belon classified the toucan among the waterfowl.

Toucan heads first appeared in Europe in the early 1500s, soon after Brazil was claimed for Portugal by Pedro Álvares Cabral. The first mention of toucans is in 1526, from the Spanish naturalist Gonzalo Fernández Oviedo de Valdés, who describes the bird as only slightly larger than a quail, but appearing much larger because of its 'beautiful, thick, vari-coloured plumage'. He also says – incorrectly – that its beak weighs more than the whole body.[12]

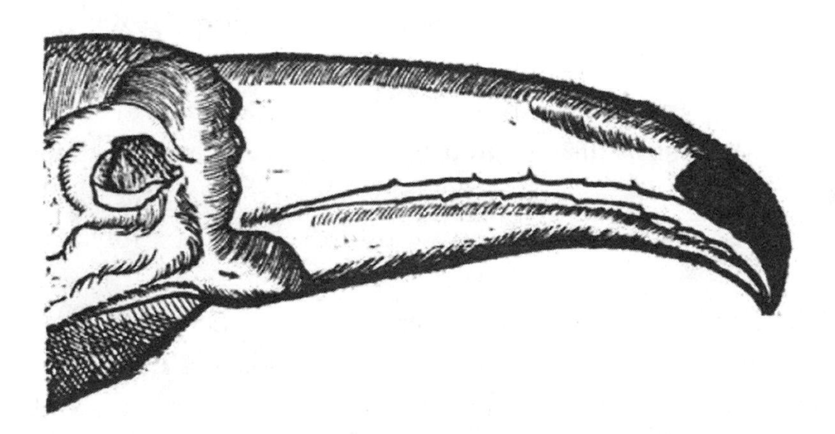

The beak of an unknown bird from the New World, illustrated and discussed by Pierre Belon in 1555. It was, of course, a toucan.

The toucan's lack of nostrils was explained by later writers, who thought they were unnecessary because the bill's jagged edges prevented the bird from closing it completely and the bird breathed through its mouth. Either that, or the toucan breathed through the bill itself.[13]

After reading Belon's book with his account of the toucan, the French Franciscan priest André Thevet, who had travelled in 'Antarctic France' (i.e. Brazil) for ten weeks in 1555 and 1556, saw an opportunity for some one-upmanship. In his own work, *Les singularitez de la France antarctique*, he describes how the toucan's plumage was used in hats and other articles of clothing, and snorts derisively at Belon's suggestion that the toucan is aquatic – 'not in my experience', he says. Finally creating an illustration, Thevet adds the nostrils that Belon failed to find. One can almost feel Thevet thinking: 'Huh, I'll show him!' But the nostrils are a joke; they are in the wrong place, and he then proceeds to shoot himself in the other foot by incorrectly drawing the toucan's toes with three pointing forward and one back, instead of two facing forward and two back. Of course, no one at the time would have known that Thevet's illustration was wrong, but as more and more specimens started to appear, he realized his mistake, and in a rehash of his illustration in 1575 the erroneous nostrils disappeared and the feet were more accurately depicted. Even so, both drawings are poor: the beak is disproportionately large and the birds seem horribly top heavy.[14]

The Toco Toucan saga encapsulates many of the issues ornithologists had with new discoveries in new lands. Inaccurate descriptions and illustrations of previously unknown birds and other exotic wildlife wormed their way insidiously into the literature and perched there, radiating misinformation.

André Thevet's top-heavy toucans. On the left, his original illustration from 1557 with the erroneous nostrils in the middle of the bill and incorrectly placed toes (courtesy of Bibliothèque nationale de France). On the right, a revised version from 1575 with the nostrils deleted and the toes better depicted (courtesy University Library of Leiden).

The toucan issue was confounded further by Belon's contemporary and competitor, the Swiss scholar Conrad Gessner, who had published his own bird encyclopedia in the same year. Gessner makes no mention of the toucan in that volume because he had not even heard of it, but, keen to rectify his omission, he decided to include an image of it in a later publication, his *Icones avium omnium* in 1560. By his own admission, however, that image was an ornithological hodge-podge: the beak belonged to a Toco Toucan, but the body was that of another toucan species based on a description by Thevet. And with three forwardly directed toes, Gessner managed to create a bird that does not exist. Yet this particular toucan survived for almost 200 years, copied by a succession of authors, until it finally expired sometime in the 1700s.[15]

Although this all seems extraordinarily messy, rather than ignoring these poorly known birds in his writings until better information was available, Belon included them specifically to trigger debate and discussion among the tiny but growing scientific community. Essentially, he was inviting colleagues to make suggestions regarding identity and affinities – a gesture of openness and collegiality that he hoped would advance the field – a new kind of relationship between birds and people.

The 1500s were a time of monumental change. On the one hand, countries like England, France, Spain and Portugal were sending ships to far-off lands on voyages of discovery. On the other, Henry VIII's break with Rome and the Reformation of English religion in 1533 resulted, through confiscation of the monasteries' monies, in a massive increase in the wealth of the royal family and its favoured associates.

For the privileged class, the Tudor Age (1485–1603) was a golden one. But for the masses, most of whom lived in the countryside, it was: 'a desperate daily struggle against hunger, squalor, disease, unemployment and death'. Throughout the 1300s plague had ravaged the human populations of Europe and North Africa – killing around 300 million people, about half the population – and outbreaks continued to occur over the next two centuries. By 1525 the population was beginning to recover. As it did, the problems for the rural poor increased. A growing population required more food, and to create more grazing for their own livestock, landlords took over the common lands upon which the poor depended. Long, hard winters and adverse summers meant uncertain harvests, and in the mid-1500s a succession of crop failures pushed the rural poor into profound poverty.[16]

Certain birds had been a problem for farmers since the beginning of agriculture, but by the 1500s the competition

El Tajo de las Figuras (Tim Birkhead).

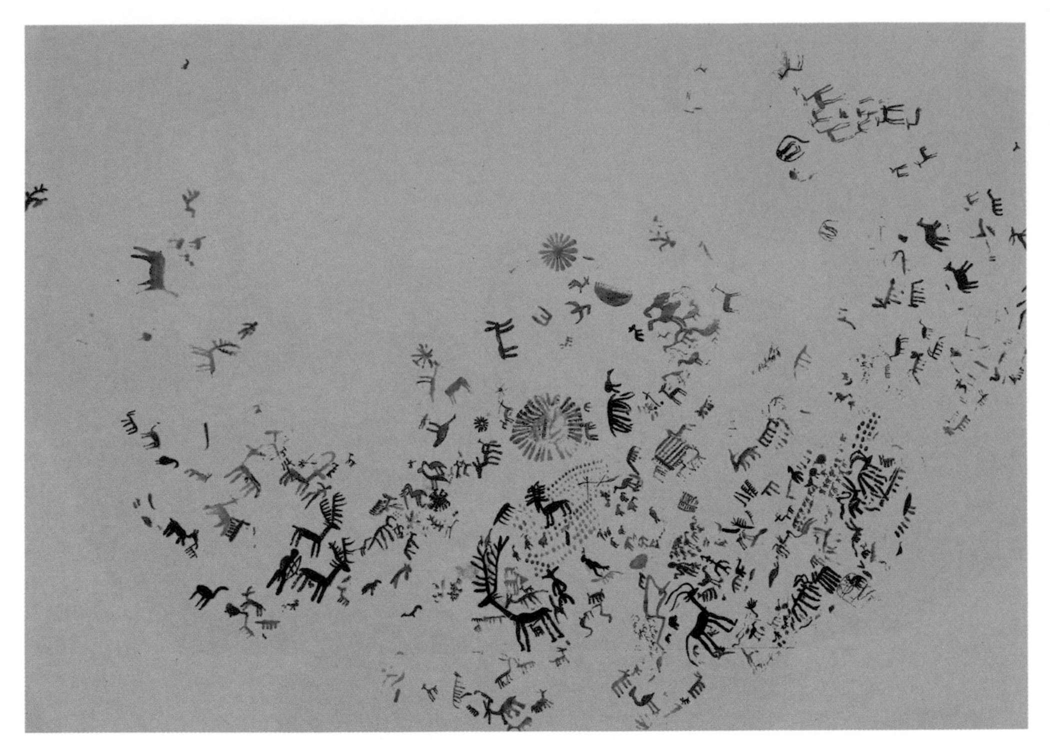

Abbe Breuil's painting of the entire painted surface of El Tajo de las Figuras (Breuil and Burkitt, 1929).

Some of the birds, mainly Great Bustards, in the El Tajo rock shelter revealed by enhanced imagery (M. Lazarich). These lie on the right hand side of the image above, running almost vertically.

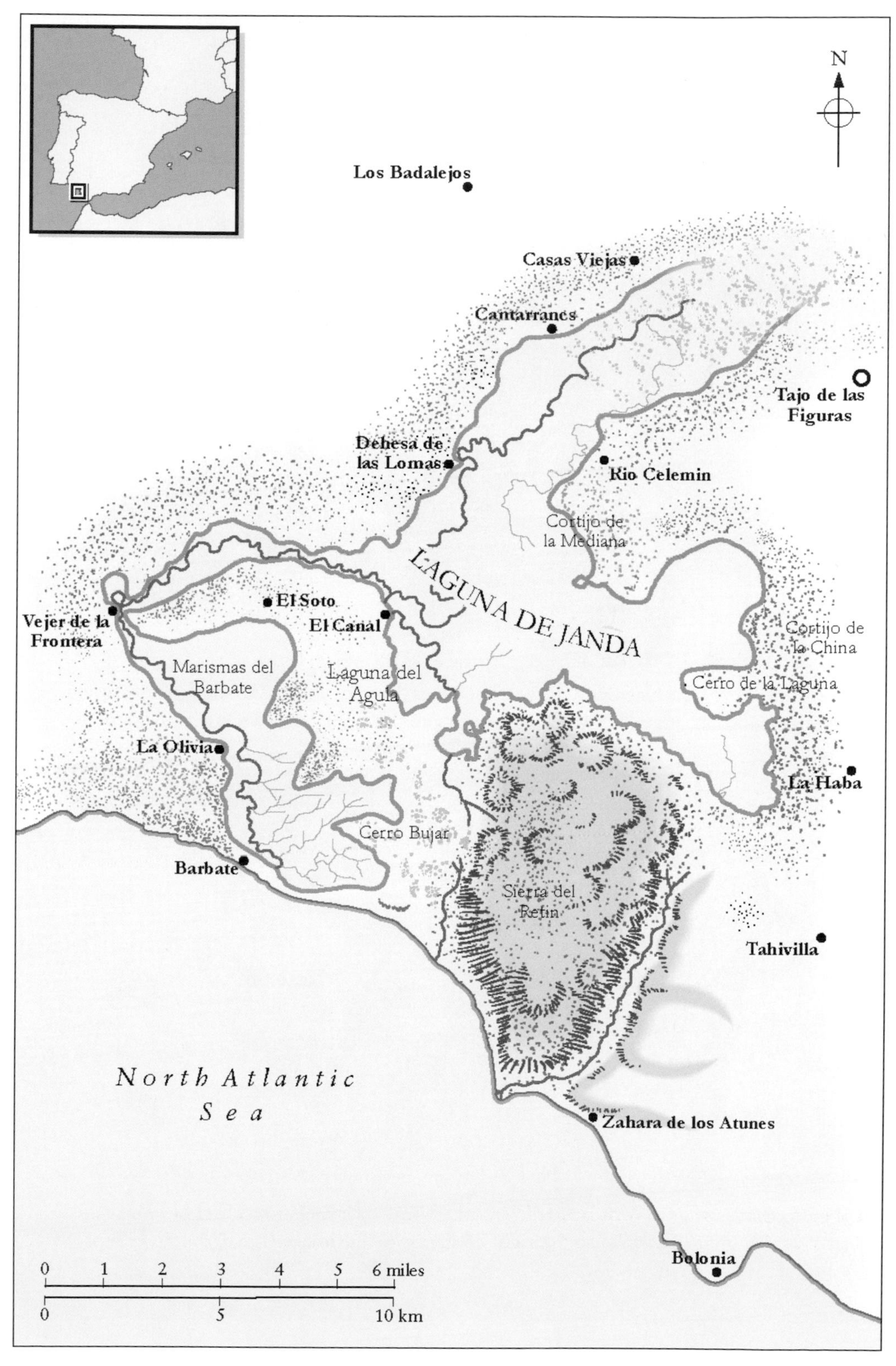

The La Janda wetland of south-west Spain as it was in the 1840s (redrawn from a map of 1841), with Tajo de las Figuras at the upper right.

Image from the tomb of Nebamun, Egypt, known as 'Fowling in the Marshes', from around 1350 BC, depicting Nebamun, his wife Hatshepsut and their daughter (Wikimedia Commons).

One of very few tomb paintings depicting Sacred Ibises in captivity, here feeding from a hand-held bowl. It is not certain that the original of this image, from the cenotaph of Alexander the Great at Medinet Madi, still exists (Bresciana, 1980).

An Egyptian fantasy of 1888: a painting by Edwin Long, 'Alethe Attendant of the Sacred Ibises in the Temple of Ibis', Memphis (Wikimedia Commons).

Roman mosaic in Thysdrus, Tunisia, third century AD, showing a decoy owl and various birds, including bee-eaters, that have been lured to their deaths (Juha Puikkonen / Alamy Stock Photo).

Roman mosaics: Barbary Partridge (*left*) from the Villa of the Aviary, third Century AD, and a possible flamingo (whose beak is far too long) from the sixth century AD (© David Tipling / naturepl.com / Premasagar via Flickr).

Two Gyrfalcons, painted by Jakob Bogdani around 1695. A white Gyr was, and still is, the ultimate falconer's bird (Wikimedia Commons).

Images from Frederick II's *De arte venandi cum avibus* (*On the Art of Hunting with Birds*); upper: presumably Frederick with falconers; lower: a naked falconer swimming to retrieve a duck killed by his falcon (Tim Birkhead).

Part of the Bayeux Tapestry showing Harold – hawk on hand – about to depart from Bosham for France (Bayeux Museum).

Part of the Bayeux Tapestry depicting Guy (on the black mule on the right), taking Harold (centre) to William, both men carrying their hawks. The lewd figures beneath Harold represent his reputation as an adulterer (Bayeux Museum).

Peregrine and a Grey Heron: 'Hawking in the Olden Time' by Edwin Landseer, mid-1800s (Wikimedia Commons).

The vanity and profligacy of hunting. In this image, 'The Meeting of the Three Living and Three Dead', three falconers (in this case a young woman and two young men) encounter three dead men. One of the hunters recoils in terror, one imagines the apparition to have been sent by God, and the third muses at the sight of decaying humanity. Death says: 'What you are, and what we are, you will be.' In other words, 'mend your ways, before it is too late'. From the Hours of Joanna I of Castile, Netherlands (Ghent?), *c.*1500 (British Library).

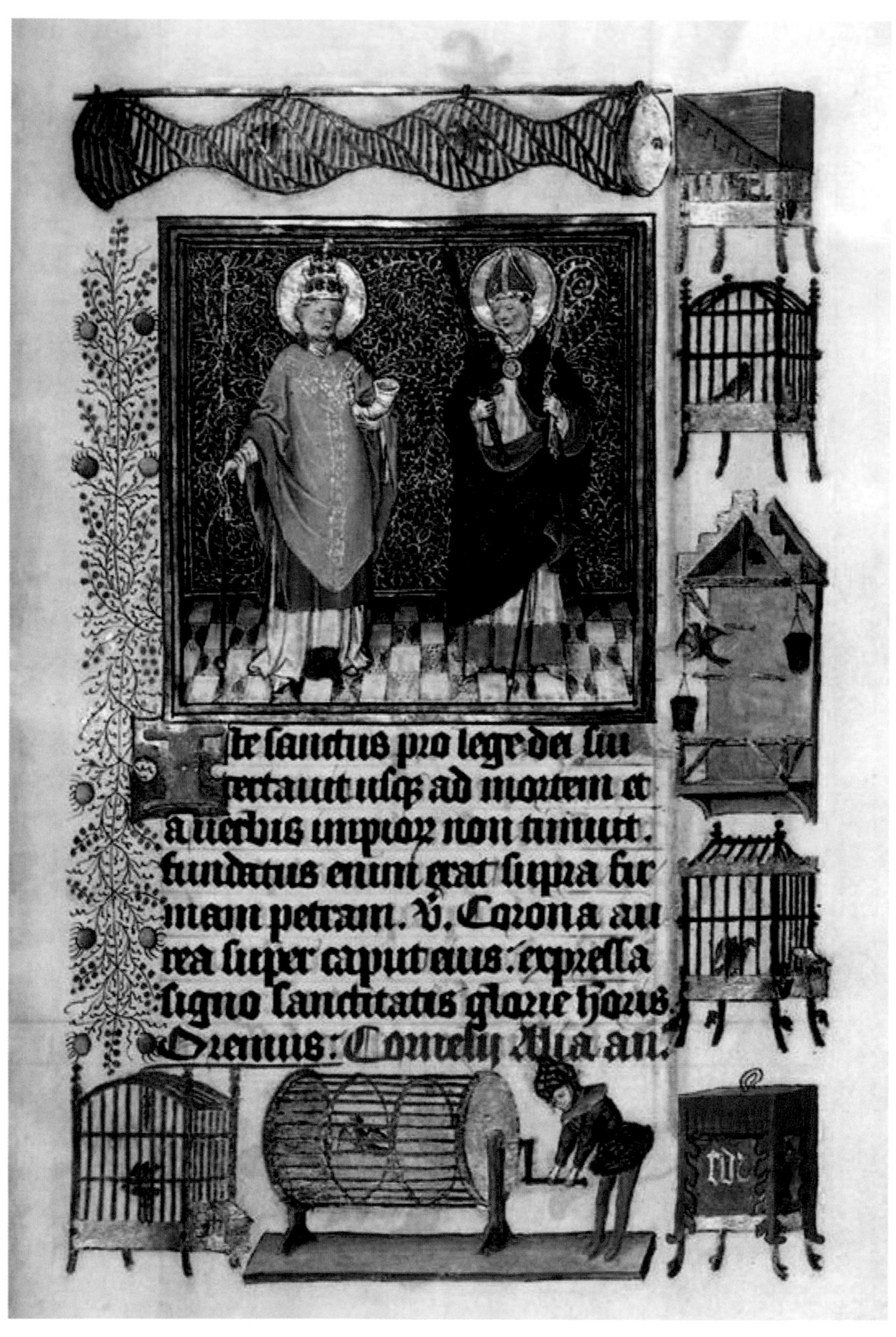

A page from Catherine of Cleve's Book of Hours (*c.*1440) with its birds and cages symbolizing Catherine's reproductive obligations. Note the obedient Goldfinch pulling a seed or water bucket on the middle right. The symbolism of the revolving cage at the bottom is a mystery (Web Gallery of Art).

A seventeenth-century painting of an Italian market stall, once attributed to Caravaggio, in the Gallery Borghese in Rome (Wikimedia Commons).

Two exquisitely executed woodpeckers, Great Spotted (*upper*) and Middle Spotted (*lower*) by Daniel Froschel (1573–1613) in 1589 at the age of sixteen. He painted for the Medici, for Aldrovandi, and may have travelled to Spain to copy Hernández's images (National Library Vienna).

Two images of a Northern Lapwing by the little-known Pisanello, painter for the Renaissance Court, from about 1430 (Wikimedia Commons).

Three of J. M. W. Turner's birds from Farnley Hall; *upper*: Grey Heron; *left*: Common Kingfisher; *right*: Peacock (Bridgeman Art Library).

Watercolours from the late 1500s of two of Aldrovandi's birds, on which the black-and-white engravings for his books were prepared; *upper*: a European Bee-eater; *lower*: a Great Curassow from neotropical America (University of Bologna).

Original, accurate, but tiny paintings of birds from north-east Brazil, by an unknown artist commissioned by Count Johan Maurits in the mid 1600s as part of his cataloguing of that country's flora and fauna (Biblioteka Jagiellonska, Krakow).

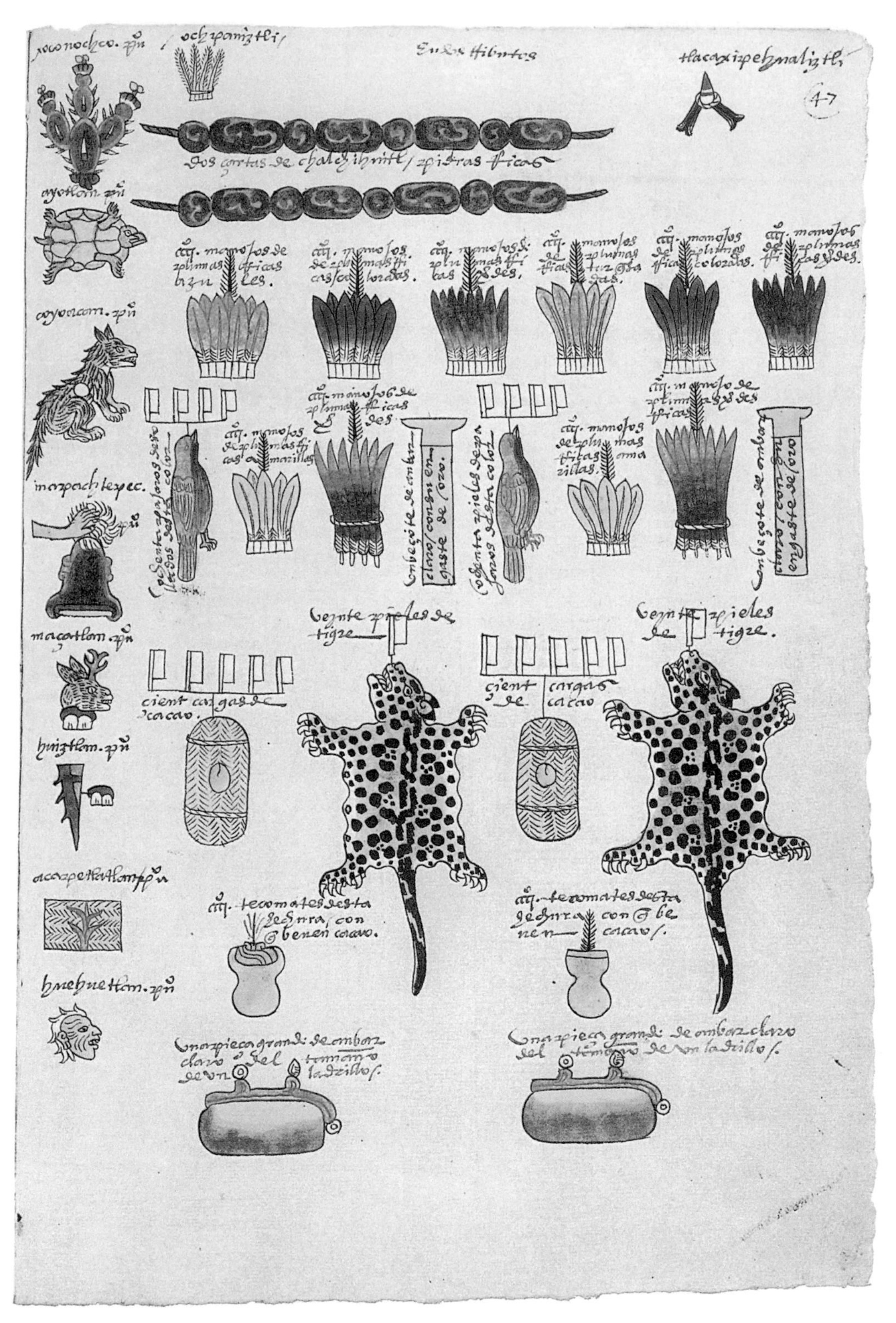

A page from the Codex Mendoza (from 1541), showing jade beads, bundles of feathers, Lovely Cotingas, jaguar skins, liquid amber (*bottom*) and packets of cacao (*left* of Jaguar pelts) (Bodleian Library, Oxford).

A modern reconstruction of a Resplendent Quetzal feather headdress, created from the plucked tail feathers of male birds (Wikimedia Commons).

Sarcophagi at Carajai, Peru, from the pre-Inca Chachapoya culture of the 1470s, some of which may have contained bird mummies (Tim Birkhead).

Watercolour paintings executed to help prepare Duke Friedrich's heavily symbolic Queen of America pageant of 1599 in the Duchy of Württemberg. The participants are bedecked with exquisite and enthralling featherworks brought from the New World (Klassik Stiftung Weimar).

Some Faroese birds painted by the nineteenth-century self-taught local artist Didrik Sørensen (Alamy Stock Photo).

A Cuckoo chick ejecting a Meadow Pipit chick from its nest; *upper*: as witnessed and drawn by Jemima Blackburn (1872); *lower*: as elaborated – with the imagined presence of a parent pipit – by John Gould in volume 3 his *Birds of Great Britain* (1873) (Alan Blackburn (*upper*) / Zoological Society of London (*lower*)).

'The Ornithologist or The Ruling Passion', depicting John Gould surrounded by his family and bird skins, as painted by John Everett Millais and exhibited at the Royal Academy in 1885. Note the Resplendent Quetzal held by the girl on the left (Wikimedia Commons).

Some of John Gould's birds; *right*: Resplendent Quetzal (Gould 1838, plate 166 – Biodiversity Heritage Library); *upper left*: Marvellous Spatuletail (from Gould 1861, plate 153 – Zoological Society of London); *lower left*: Blue-tailed Hummingbird or Long-tailed Sylph (Gould 1861, plate 172 – Zoological Society of London).

Gurney's Pitta, first described by A. O. Hume in 1875 and named for the banker and ornithologist John Henry Gurney. The painting is by William Hart – one of John Gould's colourists. Once thought to be near extinction, this endangered species is eagerly sought by twitchers visiting the Malay Peninsula (*Stray Feathers*, 1875, Zoological Society Library).

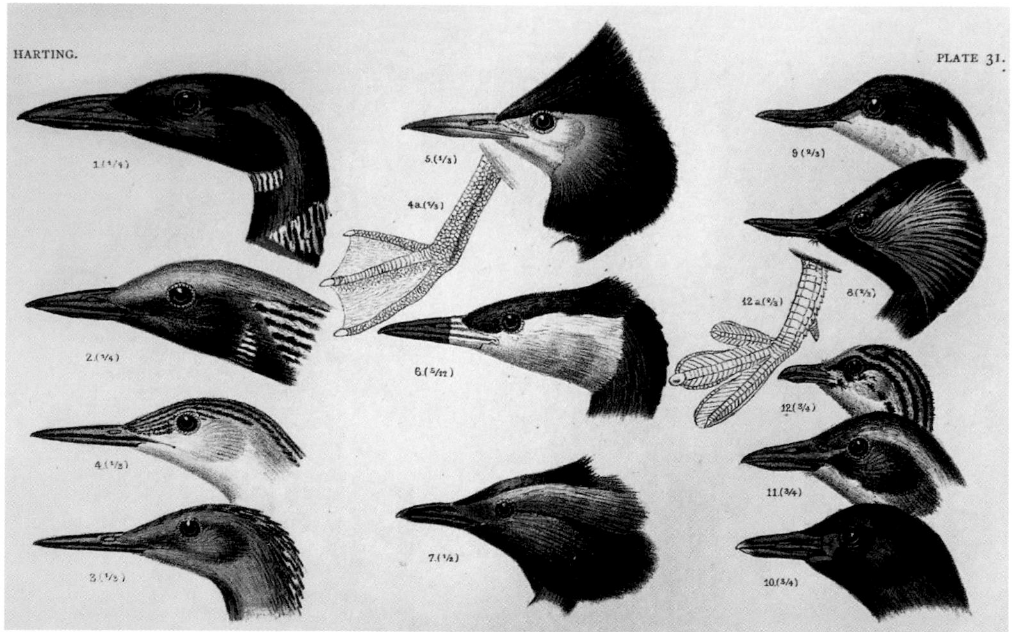

The start of the field-guide type of bird image: these are from James Edmund Harting's *Handbook of British Birds* (Harting, 1901).

Magdalena and Oskar Heinroth and some of the many birds Magdalena hand-reared, photographed by Oskar between the 1910s and 1930 (Klaus Nigge and the Staatsbibliothek zu Berlin).

Francis Lydon's images for F. O. Morris's *History of British Birds*; clockwise from top left: Common Rock Thrush, Jack Snipe, Great Northern Diver, Common Starling, White's Thrush and Golden Plover (Morris, 1850–57).

Two bird species with a polyandrous mating system; *upper*: Comb-crested Jacana by John and Elizabeth Gould; *lower*: Painted Snipe (female on right) by Henrik Grönvold (Baker, 1921) (Zoological Society of London).

Upper: Atlantic Puffin; *lower*: Common Guillemot, painted by Jemima Blackburn in the late 1800s (Alan Blackburn).

Guillemot ringing team on Skomer in 2017; *left to right*: Ed Stubbings, Jamie Thompson, Julie Riordan, Jason Moss, Elisa Miquel Riera and the author (Tim Birkhead).

Guillemots and their chicks at the right age for ringing (banding) on Skomer (Tim Birkhead).

The Guadalupe Caracara, rendered extinct in the early 1900s by collectors – painting by J. J. Audubon (plate 161 in his *Birds of America* 1827–1838) (Wikimedia Commons).

between humans and birds for food resources had become intense. It is not hard to imagine the desperation felt by those scraping a living from the land on seeing birds and other wildlife gnawing into their precious crops. This desperation resulted in a series of Vermin Acts, not so much to protect the poor but to try to create a sense of social unity and avoid the unrest that had been seen previously. In 1533 Henry VIII was about to break with the Catholic Church and launch the political and religious changes that would realign England, and he needed the country's support to do so. Ensuring that people had enough to eat was part of the plan, and in 1533 he passed the first Act of Parliament designed to protect the poor from birds.[17]

> Innumerable number of Rooks, crows and choughs do daily breed and increase throughout this realm, which do yearly destroy, devour and consume a wonderful and marvellous quantity of corn and grain of all kinds.[18]

This was the justification for an Act that stated that every English parish should have a net for catching crows and that a bounty be paid for each crow or chough killed. The church-warden was to provide the net, keep the necessary records and make the payments.

Under Elizabeth I, Henry's daughter, the war against animal competitors intensified, and the Act of 1566 'for the Preservation of Grain' identified no fewer than nineteen different birds and thirteen other animals as 'Noyfull Fowles and Vermyn'. Indeed, anything considered harmful to peoples' interests, whether it was to their cereal crops, poultry, fruit or fish, was now a legitimate target, and people were free to use 'Nettes, Engynnes and other reasonable Devyses' but not 'Handgunnes or Crossbows' to do so. Corvids, including 'choughs' (that probably included Western Jackdaws), were especial targets, as were Eurasian Bullfinches that stripped the buds from fruit trees, and House and Tree Sparrows that ate grain on the stem and also after harvest. For the rich and poor, ponds and rivers were important sources of fish – important for those days on which meat was forbidden by the Church – so the persecution of piscivorous birds like Great Cormorants, European Shags, Western Ospreys and Common Kingfishers is not unexpected. The bounties paid by the churchwarden varied according to the damage the animal was thought to inflict: a penny for every dozen heads of corvids, raptors, Green Woodpeckers, finches or kingfishers, and twelve pence for the head of a fox, badger, polecat, weasel or stoat, and two pence for that of an otter or hedgehog.

The Act was careful to state that no one should interfere with the breeding of hawks, herons, rabbits or doves, which were reserved for the pleasure of gentlemen hunters.

The effect of these Acts on British wildlife cannot be underestimated. They were equivalent to inserting a vermin-killer gene (or possibly doubling up on one already present), into the English nobility's genome. In the following three centuries the indiscriminate slaughter of thousands and thousands of birds and mammals (as well as some reptiles and amphibians) continued unabated across the British, and much of the European, countryside. Subsequent efforts to eliminate, or even reduce, the potency of the gene – or more accurately, meme – have so far failed.

The mortality imposed by this relentless persecution massively reduced the populations of some species, but far worse was the way this pursuit of predators fixed in the minds of many people the idea that certain birds and mammals are 'vermin' and could therefore be destroyed with impunity.

It was a continuation and exaggeration – but now with legal justification – of man's dominion over everything, the 'us and them' distinction between humans and non-humans, and a peculiarity of our relationship with birds and other wildlife.

The widespread, largely unthinking persecution of wildlife during the Tudor period in England reminds me of the similarly ignorant edict issued by the Chinese leader Mao Zedong in the late 1950s and early 1960s, when, as part of the Great Leap Forward, he instructed people to kill sparrows (mainly Tree Sparrows) because of the birds' predilection for grain. The people complied, of course, and the effect was an ecological disaster. The deaths of millions of sparrows that, as well as eating grain, also consumed vast quantities of crop-consuming insects, were followed by plagues of locusts that destroyed rice crops. The resulting ecological imbalance contributed to a famine in which between 15 and 45 million people died: one of the worst human-induced catastrophes ever.[19]

Sparrows, mainly House Sparrows, were persecuted in Britain from Tudor times onwards, although not to quite the same extent as in China. In part this was because of their depredation of grain, but also because of their sheer abundance. Anything that was abundant could (and can) so easily be designated as a pest, or vermin. Sparrows were subject to extreme abuse. There were 'sparrow clubs' in England and on the Continent in the 1800s and early 1900s in which the birds were trapped and sold to shooting clubs for the mere pleasure of shooting them. Today, both House Sparrow and Tree Sparrow populations in Britain are much reduced (and much missed).

Four years after the Battle of Hastings the Lestrange family arrived in England from France, taking up residence at Hunstanton on the north Norfolk coast, where they remained until the 1940s. In the 1500s Hunstanton Hall was overseen by Sir Thomas Lestrange, who had served as 'esquire of the body to Henry VIII' at the Field of the Cloth of Gold in France in 1520 (that is, he was responsible for dressing and undressing the king). The Hunstanton household accounts, maintained by Sir Thomas and his wife Lady Anne between 1519 and 1545, provide a unique insight into their relationship with birds during this period, documenting the species procured by the estate's staff, or by local fowlers, to be consumed by the family. The accounts also include the names of the wildfowlers, the delivery date, the method by which the bird was obtained (either alive, or more usually dead) and the prices the Lestrange family paid for them.[20]

Looking at those accounts today, one of their most striking aspects is the names by which the birds were known; some of which, without additional research, would be unintelligible.

'Popeler' was the medieval name for the Eurasian Spoonbill, a name that alludes to the shape of the beak. This species almost certainly bred on or near the Hunstanton estate, for the accounts refer to young birds that will have been hooked from their nests in the same manner as Grey Herons, for the table. When Henry VIII's lord chancellor, Cardinal Wolsey visited the area in 1521, he was apparently presented with three spoonbills, three bitterns, ten cygnets, twelve capons, thirteen plovers, eight pike and three tench from the Hunstanton estate – quite a feast![21]

On 20 October 1538 John Syff was paid three pence for a 'woodcock, a spowe and a cokell doke'. The Eurasian

Woodcock – a very tasty bird – is one of the few species that, after the Saxons named it (wuu-coc or wudecocc), has apparently not been lumbered with any other local names. Of the other two, cokell doke can be guessed at: cockle duck, which suggests a diving duck and is assumed to be the Common Scoter, which was abundant in the area.[22]

The spowe was thought to refer to the Eurasian Whimbrel, since the name is superficially similar to that by which this species is known in Iceland and Scandinavia: *spoi*, *spou*, *spof* and *spove*. But there's a problem, because the whimbrel, we now know, appears on the Norfolk coast only as it heads north on migration in late April and May, and then again between July and October as it flies south again. The dates in the household accounts show that spowe were obtained only between September and February, and were clearly winter visitors. After accounting for all other wading birds, my colleague Fred Cooke and I decided the only species the spowe could be is the Bar-tailed Godwit.[23]

Other birds brought to the house included Eurasian Curlew, Common Snipe, Red Knot, Northern Lapwing, European Golden Plover, dotterel, Common Redshank, stints (any small wader), teal, wigeon, Brent Goose, Common Shelduck, Common Pheasant, Grey Partridge, Great Bustard, Common Crane, Grey Heron, Common Quail, Common Blackbird, Eurasian Skylark and buntings – indicating that the family enjoyed a varied protein-rich diet – to say nothing of the porpoise and pike, and huge range of marine fish such as herring, codling, lamprey, plaice, gurnard, sea bass, skate, sprats, sole and salmon. The birds were either trapped in nets, snared in horsehair nooses or killed with a crossbow. Waterfowl were retrieved by dogs ('spannyells'), and the first bird to be killed by a gun on the Hunstanton estate, in 1533, was a moorhen.

Hunstanton's resident falconer used Northern Goshawks – in a purely utilitarian way rather than for display – to kill conies (rabbits) and partridges, and in June 1533 a falconer's Hobby took fourteen skylarks.[24]

These extraordinary household accounts provide a rare glimpse into the lives of what Daniel Gurney referred to as the 'better sort of gentry', who would have been very much aware of the quality and flavour of different birds and the prices they were prepared to pay. But it was not just flavour. The health benefits and 'digestibility' (good and bad) of different species were important, and the Lestrange family may even have had a copy of Andrew Boorde's early sixteenth-century, largely medical tract, entitled *Dyetary of Helth*, wherein the pros and cons of different birds are spelled out. I suspect also that much of this too was common knowledge among the aristocracy: 'Of all wylde foul the Feasaunt is moste beste,' whereas, 'A crane is hard of dygestyon and doth ingender evyll blood.'[25]

A subsequent, and more comprehensive book on the benefits of different foods, including a wide range of birds, is Thomas Muffett's *Health's Improvement*, written around 1595. The Grey Partridge was said to 'strengthen the retentive power . . . fattening the body and increasing lust'; the bustard provided 'rare nourishment, restoring blood and seed'; quail: 'the daintiest of meat'; lapwing: 'very sweet, delicate fine flesh'; teal: 'sweet of taste'; and snipe were said to be 'so light of digestion . . . That they agree with most men's stomachs'. The flesh of young cock-sparrows 'as well as their stones [testicles]' and brains should be served 'to such [people] as be cold of nature, and unable to *Venus* sports' – i.e. a remedy for impotence. Increased lust, lechery and the replenishment of 'seed' seem to have been high on the list of benefits of eating birds, as they probably would be today, were it true.[26]

Judging from the books of Andrew Boorde and Thomas Muffett, people (well, largely men) seemed to have been preoccupied by the effect that eating birds (and other animals) would have on their digestive system and libido. Looseness of the bowels or its opposite was a constant concern. A soup made from larks could result in diarrhoea, but eating lark flesh could cause constipation. Eating a heron 'green roasted' – that is, without prior boiling – could result in 'piles and smarting hemerrhoids'. More positively, perhaps, duck, turkey, partridge, bustard and sparrow were all thought to increase lust and 'the seed'. The testicles of different animals were said to 'work marvels in decayed bodies' by 'stirring up lust through an abundance of seed'. 'Indeed,' Muffett continues, 'when Bucks and Stags are ready for the rut, their stones and pisels [penises] are taken for the like purpose: as for the stones of young Cocks, Pheasants, Drakes, Partridges, and Sparrows, it were a world to write how highly they are esteemed.'[27]

It was but a small step for Renaissance men to move from contemplating the digestibility of particular birds or their effect on concupiscence to imagining their medicinal properties. And no wonder – disease and ill health were widespread in the 1500s and 1600s, with little in the way of effective remedies or analgesics. As the very name implies, 'smarting hemerrhoids', caused or exacerbated by constipation, could make life unbearable. The prospect of the plague, leprosy or syphilis, or conditions like 'the stone', clearly haunted peoples' lives. Most remedies were herbal; concoctions derived from a wide variety of plants, often informed by the 'doctrine of signatures' – a link between the appearance of the plant (the sign) and the disease it was intended (by God) to cure. Lungwort, for example, whose spotted leaves were thought to resemble diseased lungs, were used to treat lung disease. Sometimes it could work the other way round: people avoided, for example, eating wagtails because their ever-trembling tail reminded them of the shaking palsy – Parkinson's disease. The 'doctrine' idea had its origins with the Greek physician Dioscorides in the first century AD, and was promoted by the rather better informed Paracelsus in the decades immediately before Boorde and Muffett wrote their books.

We shouldn't be surprised, then, to find that birds, or parts of them, were used in the treatment of certain conditions. The flesh of Domestic Pigeons, bled from under the wing, 'newly killed and forthwith rosted at a blasing fire', was said to engender a 'great store of blood, recalling heat unto weak persons, clensing the kidneys [and] quickly restoring decayed spirits'. This is entirely believable, for it sounds like a nice meal.

More seriously, the flesh of chickens was thought to be a remedy for consumption (tuberculosis), pheasant for 'hectick fever', and the 'brains of cranes for haemorhoids'.

It seems extraordinary – to me, at least – that by the beginning of the Renaissance in Europe people were still treating illnesses with remedies first prescribed over 1,500 years previously. That they did so was a testament to the trust people placed in the knowledge of the ancients, to anecdotal evidence and possibly also to the placebo effect. It is also proof, if proof were needed, of the lack of progress that trust engendered. No one seems to have asked or checked whether consuming the ashes of a cuckoo actually alleviated the symptoms of 'the stone'.

In the Western world, we continue to be amazed, and sometimes dismayed, that, in some other regions, local people rely on homeopathic medicines or on remedies that are no different from those in use two or more millennia ago. In Africa, for example, over 300 bird species have been recorded for sale as traditional medicine in local markets. A recent study of traditional medicine among the Maasai people in Tanzania revealed that they used the body parts – bone, blood, beaks and so on – of over 150 bird species, including the ostrich, Kori Bustard, Carmine Bee-eater, Malachite Sunbird, honeyguides, hawks and herons.

What is remarkable about some of the Maasai remedies is how similar they are to those employed in the ancient world and in Renaissance Europe. The Maasai use the fat from a variety of birds (flamingo, ostrich, chicken) poured inside the ear for ear infections. In Europe it was heron or crane fat. For epilepsy the Maasai administer the ash from a Giant Kingfisher's beak or a widowbird's feathers. In Europe, it was a kite. In medieval and Renaissance Europe burning a bird, or a plant, and reducing it to ashes was thought to be a way of capturing its essence, and perhaps that is true also of the Maasai.

It is difficult to know whether there was any flow of medical knowledge from Europe to Tanzania, or whether this is a case of what in biology would be called 'convergent evolution' – the same features arising independently in different populations subject to similar ecological conditions.

The 'stone' which seemed so widespread in Renaissance Europe is not recorded as an illness that the Maasai suffer from, or at least, not in the study reported, but they are susceptible to malaria, typhoid, tuberculosis, syphilis, ringworm, sleeping sickness, Parkinson's disease and elephantiasis, all of which are treated in one way or another using bird parts. Westerners question the efficacy of much traditional medicine, certainly compared with our own, but the Maasai consider most traditional treatments to be 'moderately effective'.[28]

In the Western world bloodletting, cupping and the use of emetics consisting of burnt birds, were all still widely used at the time of the Scientific Revolution in the 1650s. Indeed, such medically useless remedies continued to be employed for centuries before eventually being replaced by evidence-based medicine. Even so, homeopathic remedies are popular today, including one – with no known efficacy – named Oscillococcinum, taken in France and Russia to relieve flu-like symptoms. It is made from one part duck offal to 10^{400} parts water, a dilution so extreme that none of the original ingredient remains.

The study of natural history has its roots – literally – in herbal remedies. The contrasts, then, between the inertia in medical research during the seventeenth century and the huge leaps forward in the study of natural history that triggered the Scientific Revolution are remarkable.

6. The New World of Science: Francis Willughby and John Ray Discover Birds

> The first woman to attend a meeting of the Royal Society was Margaret Cavendish, Duchess of Newcastle, in May 1667. Margaret later raised issues that have become perennial. She mocked the dry, empirical approach of the fellows, violently attacked the practice of vivisection and wondered what rational explanation could be given for women's exclusion from learned bodies.
>
> Richard Holmes (2010)

As a teenager, rather than simply listing the birds I had seen, I was filled with a sense of what some have called 'restless endeavour' – an all-consuming desire to do something creative within the field of natural history and ornithology.

With no real idea of what scientific ornithology entailed, I scanned about for a project, something to focus on, and that something turned out to be the Grey Heron: large, majestic and charismatic, and in the 1960s, a victim of pesticide poisoning.

My local patch was a small eutrophic lake surrounded by woodland in the private grounds of Farnley Hall in the valley of the River Washburn. Herons congregated here to feed at dawn and dusk, and during the rest of the day could be found in a straggling group of up to a dozen on the adjacent fields, on what in heron folklore was known as a 'standing ground'.

Hunched up with their backs to the wind, the birds resembled a little gathering of grey, shrouded monks.[1]

For hours on end and in all weathers I watched them on their standing ground, naively hoping that simply by putting in the hours I'd see something that no one else had seen, revealing the true function of the birds' behaviour. I didn't, and indeed I could hardly have chosen a more stultifying project, for herons actually do nothing on their standing grounds. Very occasionally I'd be rewarded by witnessing a scuffle with raised plumes and squawks as one bird alighted too close to another, but mostly they simply stood there. And, of course, as I eventually realized, doing nothing was the whole point: the birds were resting and the standing ground was their daytime roost.

My time was far from wasted, though, as watching herons developed my patience and honed my observational skills. It has been suggested – controversially – that to become expert in something, one has to put in 10,000 hours of effort. The time I spent watching herons at Farnley as a teenager – my apprenticeship – contributed substantially to my 10,000 hours.

While watching the herons I was occasionally challenged by the estate's gamekeeper for trespassing. Undeterred, I tried to become better at avoiding him. Eventually, our family doctor, who was keen on birds, came to our house to gently 'have a word', advising me that if I wished to continue visiting Farnley I should seek permission from the owner, Nicholas Horton-Fawkes, a descendant of the notorious Guy Fawkes. Knowing I was interested in art, the doctor also mentioned that inside the hall, in a darkened room behind shuttered windows, there was a collection of watercolours of birds by the world-renowned artist J. M. W. Turner. I could only imagine what these paintings were like, for at that time they had not been reproduced and were not accessible to the public. I longed to see them, but having failed to endear myself to Mr Fawkes, I felt unable to ask permission, either to continue watching the herons or to see the paintings. At the time, it didn't really matter, as simply knowing that Turner had been at Farnley, and that we had a mutual passion for birds and art, was sufficient.

Subsequently, thanks to the work of art historian David Hill, Turner's bird paintings were published in 1988. Opinions varied regarding their quality. My mum took one look at them and snorted derisively, saying, 'You could do better.' John Ruskin, on the other hand, thought them extraordinary, and was especially enchanted by Turner's exquisitely rendered Common Kingfisher.[2]

Almost fifty years were to pass before, in 2018, I found myself driving past the Farnley estate once more. My curiosity got the better of me and, trespassing once again, I walked down to the lake, only to find it much overgrown, but still with herons fishing along its edge. Inspired, I looked on the internet and found that the estate, now run by Nicholas Horton-Fawkes's son, one Guy Fawkes, could be visited. A few weeks later, with Guy as my guide, I had the pleasure of finally seeing Turner's paintings. I was also shown the Fawkes family's 'Ornithological Collection', an album comprising three bulky volumes, for which Turner had created many of the bird paintings. Compiled during the early 1800s, the album was the brainchild of Walter Fawkes, whose love of birds had been inspired as a boy by reading *The Ornithology of Francis Willughby*.[3]

Turner, of course, is best known as a painter of majestic landscapes rather than birds. I can imagine him as a house guest at Farnley being encouraged to contribute to the bird album, sometimes not very successfully, as with some of the smaller birds, but occasionally hitting the mark, not only as he did with the Common Kingfisher that Ruskin so admired but also with the vibrant Grey Heron, with its gleaming eye and small fish clasped in its beak. For me, this is Turner's ornithological masterpiece.

On estates like Farnley, wealthy landowners once vigorously protected their herons. During the Middle Ages this was done to ensure an adequate supply of falcon quarry, but later it was to provide a supply of young herons for the table. When I first read about this I found it hard to believe that herons were farmed for their fledglings, in part because I had previously read W. H. Hudson's account of eating a heron that sounded truly disgusting. However, if young herons were 'stewed' – that is, kept in pens and fed on ox liver and curds – to fatten them up, and then roasted with 'swyne fat and flavoured with ginger and pepper, they were said to taste somewhere between hare and goose'.[4]

The bird book that inspired Walter Fawkes was written by two groundbreaking ornithologists of the Renaissance whose early research experiences included witnessing a harvest of young herons. During their Grand Tour, Francis Willughby, John Ray and their two travelling companions arrived at the village of Sevenhuis, in the Netherlands, on the evening of 5 June 1663, where a 'remarkable grove' held a vast colony of Grey Herons, Eurasian Spoonbills, Great Cormorants and Black-crowned Night Herons. The next day they went out in a boat to watch the harvest, noting: 'They shake down the young ones by a hook fastened to a long pole.' Each year since the 1400s the landowner at Sevenhuis had managed, protected and harvested several thousand young birds produced in the 'heron wood'. Willughby and Ray, whose research was about to change the face of ornithology forever, used the opportunity to collect a few heron eggs, and dissect some specimens.[5]

The Scientific Revolution of the Renaissance comprised two strands: the rich interwoven threads of astronomy and mathematics, pioneered by Copernicus, Galileo and Isaac Newton, and the subtler filaments of natural history forged by Willughby and Ray.

I lived, metaphorically, with John Ray and Francis Willughby for over a decade in my research, trying to get inside their heads to better understand the seismic shift their studies of birds generated.[6] Willughby, from an aristocratic, landowning family, went up to Cambridge just before his seventeenth birthday, where he was tutored by Ray, the son of a village blacksmith and a herb-woman and eight years Willughby's senior. Both were brilliant, albeit in different ways, and both were fervent adherents of the scientific study of natural history. Together they searched for plants in the Cambridgeshire countryside, later extending their horseback journeys further afield. Returning from a tour of Wales in June 1662, Willughby suggested to Ray that they consolidate their interests by overhauling the whole of natural history. Independently wealthy and committed to a life of study, the intellectually exuberant Willughby was frustrated by existing accounts of the natural world. Reorganizing the entirety of natural history knowledge was wildly ambitious, but both men were driven by their enthusiasm for 'the new philosophy' – later referred to as the Scientific Revolution they had helped to create in Cambridge. Ray's interest was primarily in plants, and so they decided to divide up the work: Willughby on animals, Ray on 'vegetables'. In fact, they worked together on most projects, Willughby's agile mind beautifully balanced by Ray's steadying influence creating an intellectual partnership like few others.[7]

They decided to start with birds, and over the next few years accumulated notes and specimens in preparation for their encyclopedia of ornithology. The previous encyclopedias of birds comprised a messy mix of fact and fantasy, mythology and genuine natural history, all overlain with a gloss of supposedly irrefutable classical knowledge. In contrast, the 'new philosophy' championed first-hand, objective knowledge, captured in the motto of Britain's Royal Society, of which both Ray and Willughby were early members, *Nullius in verba* – take nobody's word for it.

Their main goal was an objective 'arrangement' – or classification – of birds. Although they didn't say so directly, they were interested in God's plan – the long-established idea that species had not been created at random, but rather that God had had some grand design in mind, which, once resolved, would provide a firm foundation from which the study of birds could grow.

It meant describing *all* known birds. This was a formidable task, but even more formidable had they been aware of just how many bird species – almost 11,000 in total – we now know exist. Because so little of the world had been explored by the mid-1600s, Willughby and Ray thought that there might be around 500 different bird species to deal with, a number they considered manageable.[8]

During their Grand Tour across Europe in 1663–4 the two men searched assiduously for birds, acquiring specimens, describing them inside and out and identifying what Willughby called their 'characteristic marks' – an exercise in seeing similarities and differences. What distinguished, for example, the Common Buzzard from the very similar Honey Buzzard? This approach, at which Willughby excelled, involved looking at birds and not relying on what previous authors had written. It was a pragmatic leap forward: the systematic collection of information following a prescribed methodology – of their own making – was a conceptual leap forward too. They shot birds, they bought birds from food markets in Italy and Spain, they were sent birds by friends and they examined specimens and illustrations in the private collections of friends and colleagues.

The result of their combined efforts was a practical, common-sense classification, based on habitat and anatomy – a more sophisticated version of what Aristotle had once suggested. Willughby and Ray categorized birds as either aquatic or terrestrial, and then further divided them according to the structure of their beaks and feet: hooked beaks, long beaks, fine beaks, broad beaks; webbed or unwebbed feet, or birds with fused or unfused toes. Not only was theirs an arrangement that surpassed all previous attempts, it was one that stood the test of time. A century later, Carl Linnaeus leant heavily on Willughby and Ray's work for his own classification.

Their considerable efforts appeared in their groundbreaking book whose title was a tribute to Willughby. He had died at the age of just thirty-six in 1672 and had financed the entire venture. After Willughby's death, Ray continued to live in the house with Willughby's widow Emma and their children while he collated their numerous notes and started to write the book. The *Ornithology* was published first in Latin in

1676, and two years later in English, and remained the standard ornithology text for the next two centuries.[9]

British birds were well covered in the *Ornithology*, Continental ones slightly less so, simply because Willughby and Ray had had less opportunity to see them. But it was birds elsewhere in the world that posed the greatest challenge. In their determination to be comprehensive, Willughby and Ray had pored over the published accounts of birds seen by others in the Americas, Africa and the Far East, and struggled to separate fact from fiction in what were often carelessly written and poorly illustrated accounts.

The relationship Willughby and Ray had with birds was an intimate one, primarily because identifying their distinguishing marks meant examining birds in the hand. There was no other way; there were no binoculars and no field guides. To find the distinguishing marks they had to look closely at the plumage, the beak, the feet, to say nothing of the guts and gonads. But I'm not sure it was an affectionate relationship. John Ray, whose primary interest was plants, had once waxed lyrically about the flowers he encountered on a walk near Cambridge, but in all my readings of Willughby's works – and there aren't that many because most have been lost – I failed to find any hint of an affection for birds. In this respect, Willughby was perhaps the archetypal, objective scientist, even though in his private life and with others he was warm and caring and, notwithstanding his privileged social position, considerate to the poor who came begging at his door.[10]

For Ray, being comprehensive meant knowing everything that had previously been written about birds. As he prepared the *Ornithology* after Willughby's death, he read widely, systematically and carefully. Of the several existing encyclopedias of birds the potentially most useful was Ulisse Aldrovandi's three-volume *Ornithologiae*, published between 1599 and 1603. Written in an age when the number of pages spoke louder than their content, Aldrovandi exulted in completeness, albeit in a rather uncritical and verbose manner. Crucially, though, he had seen birds that Willughby and Ray had not, and they were keen to include them in their own book. In some ways, the most useful aspect of Aldrovandi's monumental efforts was the images. Being wealthy, and living in Italy, the European home of artistic expertise, Aldrovandi had access to excellent artists, whom he commissioned to create coloured images of the birds on which the woodcuts for his book would be based. Almost inevitably the paintings – which Willughby and Ray saw while in Bologna in 1664 – are far superior to the resulting black-and-white woodcuts. Aldrovandi's exotic birds – most of which Willughby and Ray never saw dead or alive – included several from the Cape of Good Hope in South Africa, from where there had been trade with Europe since the early 1500s. These included the tiny Pin-tailed and Paradise Whydahs, whose males' long tails made them popular in the Medici family's aviaries.[11] That these little birds were able to survive the long boat journeys to Europe says more about their constitution than it does about those responsible for their safe passage. Most successful imports were hard-billed birds, or seed-eaters (rather than soft-bills that fed on insects or fruit), and they were often birds that in the wild lived in relatively arid habitats. The Zebra Finch is the exemplar, imported into Europe from Australia in the early 1800s and able to survive for months on end with no water whatsoever (which it obtains indirectly by extracting metabolic water from the dry seeds it consumes). It wasn't until husbandry techniques improved that so-called soft-billed birds like toucans could be kept alive on long sea-journeys.[12]

One exotic bird that did survive the long sea voyage from south of the equator to Europe, and featured in the *Ornithology*, was the Dodo. This species, endemic to the island of Mauritius in the southern Indian Ocean, has been extinct since the late 1600s. Yet no bird so poorly known has had so much written about it. Ray's information came mainly from Carolus Clusius's *Exoticorum libri decem* – an overview of the world's exotic fauna and flora published in 1605. But the illustration Ray used in the *Ornithology* is not Clusius's rather ropey rendering, but instead is a copy of an image by the Dutch physician and naturalist Jacob Bontius, created in 1631 for his *Historia naturalis et Medicae Indiae orientalis*, but it is possible that his image had been cribbed from someone else. In fact, it would have been better if Ray had used Clusius's image, which he had copied from a sketch by Cornelius van Neck, who had seen living Dodos during his voyage to Mauritius in 1598. The several squat, dumpy Dodos painted by Roelant Savery between 1611 and 1626 are responsible for what is now the most commonly published, but specious, image of the Dodo in life. It is a misconception reinforced by Sir John Tenniel's 1865 illustration – itself based on Savery's – of the Dodo in *Alice's Adventures in Wonderland*. As van Neck and Clusius indicated, the living Dodo was a much more streamlined, agile bird than most images would suggest.[13]

Until the first genuine Dodo remains – fossils – were found in 1865, many considered it a mythical, and certainly a fantastical species. Not John Ray, however, for, as he says, soon

Three images of the Dodo. Upper left: that from Carolus Clusius (1605); upper right: that from Adriaen van de Venne (1626); lower: Ray's copy of Jacob Bontius's image from 1631.

after he had written his account of the Dodo for the *Ornithology*: 'I hapned to see in the house of Peter Pawius . . . professor of Physic in the University of Leiden, a Dodo leg cut off at the knee, lately brought out of Mauritius,' which he duly described. He and Willughby had also 'seen this bird, or its skin stuft in Tradescant's cabinet' in London. This particular Dodo may have been the same bird previously observed alive by the English theologian and historian Sir Hamon L'Estrange (a member of the Norfolk family mentioned previously) in London in 1638. He wrote:

> It was kept in a chamber, and was . . . somewhat biggere than the largest Turkey Cock, and so legged and footed, but stouter and thicker, and of a more erect shape . . . The keeper called it a Dodo; and in the ende of a chymney in the chamber there lay a heape of large pebble-stones, whereof hee gave it many in our sight, some as bigg as nutmegs, and the keeper told us shee eats them (conducing to digestion), and though I remember not how farr the keeper was questioned therein, yet I am confident that afterwards shee cast them all againe.[14]

Sir Hamon's account is the only 'irrefutable evidence that a Dodo reached Europe alive'. Once dead, this particular individual must have been acquired by John Tradescant the Younger, Charles I's gardener, who established a museum – known as the Ark – in London, and whose contents were eventually transferred to Elias Ashmole for his museum in Oxford. Until the mid-1800s, the preservation of bird skins was very difficult, and by the late 1700s the Ashmolean Dodo had deteriorated to such an extent that only the head and one foot now survive, properly preserved and resting safely in the Oxford University Museum of Natural History.[15]

The Dodo was flightless, so Ray placed it alongside the Common Ostrich and cassowary (both of which he had seen in Charles II's menagerie in St James's Park in London). He thus aligned the Dodo with 'the greatest land-birds, of a peculiar kind by themselves, which by reason of the bulk of their bodies, and smallness of their wings cannot fly, but only walk'. There seemed to be a further clue to the Dodo's affinities, for one of its attributes mentioned by Sir Hamon, and by several of those who had killed and eaten it, was the large

Killing Dodos on Mauritius in 1602, from Soeteboom (1648).

stones found in its stomach (gizzard). Clusius's illustration of the bird is accompanied by one such stone (which, if you didn't know, you might think was a misshaped egg). As Ray says: these stones are 'not bred [generated] there as the common people and seamen fancy, but swallowed by the bird; as though by this mark also Nature would manifest, that these fowl are of the ostrich kind in that they swallow hard things, though they do not digest them'. Gastroliths, as these are technically known, can be found in all birds that grind up food in their gizzard, but they also occur in reptiles, including dinosaurs, and some fish. In Ray's day gastroliths were thought to possess magical or medicinal properties. Subsequent research revealed that, far from being a relative of the ostrich, the Dodo was, in fact, a kind of gigantic pigeon.[16]

Compared with the tiny amount of first-hand information Ray had available for the Dodo, a wealth of detail existed for the birds from the New World: Mexico, Meso-America and the eastern seaboard of South America. Following Columbus's arrival in the Caribbean in 1492, the discovery of gold precipitated a wholesale invasion of the Americas in the 1500s. In 1519 Hernán Cortés and a small group of conquistadors landed on Mexico's Yucatan peninsula, and within two years had destroyed the Aztec Empire. By 1550 Spain controlled Mexico, the West Indies and Central America, and by the 1570s Peru as well. The gold and silver, together with timber, tobacco, cacao and sugar obtained from the New World launched Spain's 'golden era' between 1580 and 1680.

Birds, or bits of birds, were among the many commodities shipped back to Spain by the conquerors. In Mexico, the first conquistadors were enchanted by a small red bird they called *cardinalitus* – 'little cardinal' – the Northern Cardinal, because its perky red crest reminded them of the hat worn by cardinals. It may well have been among the many birds kept in the aviaries of the Aztec pleasure palaces that existed before the arrival of Spanish invaders. The Cardinal soon became popular in Europe too, and by the time Willughby and Ray were writing their *Ornithology* this bird was also being shipped to Europe from the Colony of Virginia in what was to become the USA, 'whence, and from its rare singing, it is called the Virginian Nightingale'. An ideal cage bird, the Cardinal was 'a lovely scarlet' with a very mobile 'towring crest'. On seeing its image in a mirror the Cardinal 'hath many strange gesticulations, making a hissing noise, lowring its crest, setting up its tail after the manner of the peacock, shaking its wings [and] striking at the looking glass'.[17]

In their domination of the Americas and their relentless quest for resources and information, the Spanish recognized the value of documenting everything they encountered and very quickly included historiographers, notaries, lawyers, scholars and scribes among the entourages of their explorers.

Three exotic birds depicted in Ray (1678). An unspecified hummingbird (left), a Northern Cardinal, aka the Virginian Nightingale (centre) from the New World and a Pin-tailed Whydah (right) from southern Africa.

Prior to the Spanish invasion, the Aztecs and other indigenous peoples were already in the habit of recording their own pictogram 'histories' in codices, simple books whose pages were made either from animal skins or from plant fibres coated with whitewash.

The Spanish too used codices to record what they found in the New World, and of the many that have been preserved, one of the best known is the Florentine Codex created by the Franciscan friar Bernardino de Sahagún in the 1560s and 70s. For the Spanish, power and religion were inextricably intertwined, so as well as appropriating material resources and local knowledge, the conquistadors also wanted the souls of the indigenous peoples. For that reason, establishing the Catholic Church in New Spain was a top priority. Evangelizing the Aztecs, however, necessitated being able to communicate with them, and that meant learning their Nahua language, and trying to appreciate at least something of their way of life and beliefs. Having learned Nahuatl, Sahagún systematically interviewed men and women to document their world in the most wonderfully non-judgemental manner. The outcome of his efforts was the 2,400-page Florentine Codex, a bilingual (Spanish and Nahuatl) account of Aztec culture, history and rituals.

Remarkably, this codex also included the Aztecs' own version of their conquest by the Spanish, which so easily might have alerted the Inquisition (already established in Mexico by 1570) to what they would have considered Sahagún's inappropriately sympathetic attitude to the indigenous people.[18]

A scholar who benefited from Sahagún's research was Francisco Hernández, who arrived in Mexico some forty years later. Originally from Toledo, Hernández made his name initially through a scholarly translation of Pliny's *Natural History*, which, together with his considerable knowledge of medicinal plants, saw him appointed as Philip II of Spain's personal physician in 1565. Desperate to undertake an expedition to New Spain, Hernández gently but persistently badgered the king until in 1570, during a lull in Spain's overseas wars, he was given leave – and extremely generous funding – to command his own expedition. Philip II was passionate about science, and the explicit aim of this, Spain's first 'scientific' expedition, was that Hernández should 'make a history of those natural things and their uses to physicians, surgeons, herbalists, Indians and other *personas curiosas* who appear to be knowledgeable in these matters'.[19]

Hernández, relatively old at fifty-five when he set out, was amazed by Mexico:

> What shall I say of the wonderful natural properties of so many plants, animals and minerals; of such different languages . . . of such variety of customs and rituals . . . of the clothing they wear, the ways in which they decorate and ornament themselves, which human understanding is barely capable of grasping . . .

In the seven years he spent in Mexico, Hernández diligently and methodically collected material and information;

much of it, as Sahagún had done, from interviews with the indigenous Nahua people, with whom – despite the ravages caused by Cortés and his followers – he often formed a close and empathetic working relationship. Like Willughby and Ray, his ornithological successors, Hernández also valued first-hand knowledge.

In between describing, collecting and commissioning illustrations of what he encountered during his time in Mexico, Hernández was also witness to the appalling mortality inflicted on the Nahua people by the *cocoliztli* (pestilence). This typhoid-like enteric fever, almost certainly transmitted by their Spanish conquerors, killed over 80 per cent of the indigenous population. Thinking as a scientist as much as a medic, Hernández recorded the symptoms and conducted autopsies of the epidemic's victims in an effort to better understand it. His empathy for the local people saw him later accused of becoming too close to them and failing to 'convert them to European ways'.[20]

With little sign of any return on his substantial investment, Philip II started to press Hernández for results. In 1576, after numerous delays, Hernández sent the king sixteen volumes containing descriptions of over 3,000 plants, 400 animals (including 230 birds) and 35 minerals, together with 4,000 colour illustrations created by indigenous artists. Philip was taken aback. Hernández had gone far beyond his original remit, supplying much more than had been originally requested. Adding insult to injury, Hernández's philanthropic attitude towards the Nahua people seemed almost subversive. Realizing that he might cause trouble back in Spain, Philip delayed Hernández's return, and decided not to publish his voluminous works. Instead, they were laid aside until 1580, when Philip instructed his new personal physician, the

Italian Nardo Recchi, to make a summary of the 'useful' information they contained. In a recent reappraisal of Hernández's work, the historian Leoncio López-Ocón comments that in creating his summary for Philip II, Recchi 'considered all the natural history as useless, as is usual in the medical profession' and omitted much of it.[21]

Hernández was allowed to return to Spain in 1577 and was devastated by Philip's actions, frustrated and disappointed that his work remained unpublished. Later, he was appalled by Recchi's summary of his work and, after a decade of despondency, died in 1587. Research by historians of science has since identified the several reasons why Philip II avoided publishing Hernández's work. First, the manuscripts were not well organized and the effort (and cost) of arranging the material into a coherent whole would have been immense. What's more, as a result of yet another war, the royal treasury was desperately short of funds. There were more sinister reasons too. Hernández's 'religious orthodoxy was suspect' since 'his friends (and quite possibly his family) included crypto-Jews and Erasmianists' – followers of an intellectual form of spiritual Christianity abhorred by the Inquisition. Moreover, by presuming to name Mexico's animals and plants, Hernández had 'encroached on God's prerogative', for in the Judaeo-Christian belief system, God alone had the authority to name – and hence create – living things. This on its own was sufficient reason not to publish Hernández's work.[22]

Historians now recognize that, had Hernández's work been published, its originality and breadth would have placed it on a par with Leonhart Fuchs's great herbal of 1542, and Vesalius's anatomy of 1543.[23] Instead, his manuscripts sat more or less ignored in Philip II's vast and forbidding

Escorial Palace for over a century. In 1605 the librarian there wrote:

> [The library] has a curiosity of great value . . . This is the history of all the animals and plants that could be seen in the West Indies, painted in their native colours . . . the very beautiful plumage of so many different birds, the feet, the beak . . . It is something that offers great delight and variety to those who look at; and no small profit to those whose task is to consider nature, and that which God created as medicine for man.[24]

Copies of Recchi's redaction leaked out of Spain and extracts of it were further copied and published, first by the Dominican friar Francisco Ximénes in 1615, and a few years later in 1635 by the Spanish Jesuit Eusebio Nieremberg, whose version, named *Historia Natura*, was what Willughby owned and Ray used when writing the *Ornithology*. Finally, in 1648–52, the Italian scholarly society Accademia dei Lincei published its own version of Recchi's summary.[25]

Hernández's original volumes in the Escorial library were lost for ever in a fire that lasted five full days in 1671.

Neither Recchi's version, nor the various copies contained many images of birds, and the few there are, are not well executed. Worse still, the birds have only Nahuatl names and their identity is uncertain.

Willughby and Ray were probably unaware of just how much was lost when Hernández's manuscripts went up in flames. Understandably, they may also have assumed that Nieremberg's account provided an accurate picture of Hernández's efforts. But Nieremberg's ornithology was vague, too vague for Ray, who felt obliged to relegate it to an appendix in the *Ornithology* that he labelled 'Such birds as we suspect for fabulous [mythical] or such as are too briefly and unaccurately [*sic*] described to give us a full and sufficient knowledge of them, taken out of Franc. Hernández especially.'[26]

As Charles Raven said, 'for Ray, who had always insisted on the need for personal knowledge of the living subject, this dealing with inadequate and often unreliable stories at second hand, with little means save his own insight for discriminating between them, must have been both arduous and distasteful'.[27]

Of all the birds described by Hernández (according to Nieremberg), and reported by Ray, the Resplendent Quetzal receives the most attention. It is described in so much detail, in fact, that I am surprised Ray doubted its authenticity, although perhaps with no accompanying image, he decided to err on the side of caution and keep it in his 'fabulous' category. Anyone lucky enough to have seen a Resplendent Quetzal in the wild will know they are indeed fabulous (in the modern sense of the word), and the space Hernández allocated to it reflects the extraordinary status with which it was held in indigenous Mexican culture. The bird's back, head and upper breast plumage gleam iridescent grass green, and it has a red belly, white under-tail coverts and, in the male, two elongated tail streamers that undulate in an almost surreal manner when the bird is in flight.

Quoting Hernández (Nieremberg), Ray wrote:

> Its feathers have made the *Quetzaltototl* [Resplendent Quetzal] more precious than gold . . . The feathers of this bird are highly esteemed among the *Indians*, and preferred even

> before God itself; the longer ones for crests, and other ornaments both of the head and whole body, both for war and peace: but the rest for setting in feather-works, and composing the figures of saints and other things; which they are so skilful in doing, as to not to fall short of the most artificial pictures drawn in colours.[28]

As the historian Marcy Norton has said, for the Aztecs, 'quetzal feathers were above all the quintessence of both feathers and the brilliant glittering greenness that not only symbolized but also emanated generative fecund life. The feathers reinforced their association with wind (a vehicle for life-giving rains), movement (the essence of animation) and brilliance (the symbol of new growth).'

She also reports how, during the festival of Izcalli, the fire god's form, or *teixiptla*, was created with a mask of green stones adorned with a crown of quetzal feathers that 'spread out . . . on each side . . . like . . . horns . . . and they laid over him a cape . . . made entirely of quetzal feathers. The wind penetrated it, making it as it were, stir upwards so that it glittered and gleamed.'[29]

The fact that the Nahua people had specific terms for different patches of quetzal plumage on the bird's body signifies – like the many Yupik and Inuit words for snow – just how important these were for them. No one killed quetzals; the birds were trapped using birdlime and then released so that their regrown plumes could be taken another day. Hernández describes how, after being captured, quetzals 'remain still and quiet, not struggling at all', and how they are so in love with their own beauty that they choose to be captured rather than to do anything that may 'deface or prejudice' their feathers.[30]

Frustrated by the lack of clarity in this New World ornithology, Ray was right to be guarded in writing about birds he had not seen, or whose descriptions seemed too bizarre to be believable. He did his best, and cautioned his readers where necessary. However, it is not clear whether he recognized the uncertainty surrounding the descriptions of these New World birds. Unlike the information for familiar British or European birds that he and Willughby had themselves obtained, information about exotic birds had invariably journeyed through the successive mouths, minds and pens of many different individuals, no more so than the description of the Resplendent Quetzal.[31]

Willughby and Ray had a second, and apparently more dependable, source of information on New World ornithology to complete their list of all known birds. This was the account by two naturalists, Wilhelm Pies (better known by his Latinized name, Piso), a Dutch naturalist and physician of German descent, and the German naturalist George Marcgrave, both of whom provided detailed descriptions and illustrations of the numerous animal and plant species they encountered in the north-east corner of Brazil in the 1630s and 40s. Their *Historia naturalis Brasiliae*, published in 1648, was commissioned by the exceptional Johan Maurits, count of Nassau and governor of Brazil between 1636 and 1644. This region of Brazil was appropriated by the Dutch East India Company in 1630 and, as with all colonial expansion, the objective was the acquisition of resources, in this case mainly sugar and Brazil-wood, extracted with the labour of slaves shipped from Africa. The riches accruing from sugar and timber were offset, however, by the huge cost of protecting and administering the new colony, and in 1637 Maurits was brought in to oversee a new style of governance. Subsequently referred to as the 'Humanist Prince', Maurits not only did what was required but also extended the new colony while at the same time actively supporting the arts and sciences. The result was a 'rich portfolio' of seventeenth-century Brazil, covering its 'botany, zoology, diseases, local medicines, ethnology, astronomy, topography, scenery and native peoples'.[32]

Piso and Marcgrave were part of the Maurits entourage, beginning their documentation of the area's natural history in 1637. Maurits employed, sometimes at his own expense, six Dutch artists to provide illustrations of the flora, fauna and people, and it was their paintings (executed in both oils and watercolours) on which the woodcuts that illustrated the *Historia naturalis Brasiliae* were based. The original paintings, comprising a curious mix of good, bad and unattractive, are often too tiny to allow those responsible for creating the woodcuts to do a really good job.[33]

In contrast to the work attributed to Hernández that Ray relegated to an appendix, he seemed happy to incorporate Piso and Marcgrave's material – notwithstanding its incomplete descriptions, less than perfect images, and local Tupi

Three New World birds: Squirrel Cuckoo (left) Boat-billed Flycatcher (centre) and Guira Cuckoo (right), from Piso and Marcgrave's *Historia Naturalis Brasiliae* (1648).

names that to the Europeans were often unpronounceable – into the main body of the *Ornithology*.

For the Aztecs in Mexico, the Tupi in Brazil, the Incas in Peru – in fact, for many indigenous people the world over, including in Europe – birds and their feathers were a vital part of their cultures.

Featherworks have been found in elite ancient Peruvian burial sites dating back to 3000–2000 BC, and, as is clear from wall paintings found at the archaeological sites of Teotihuacán (AD 500–550) and Bonampak (AD 700), Mexican and Mayan nobles, respectively, had been using feathers in headdresses to signify their status for centuries.[34]

The bright feathers from birds like the quetzal, the Lovely Cotinga, macaws, tanagers and occasionally even tiny hummingbirds were important items of both trade and the tribute payments the Aztecs demanded from the communities they conquered. These are all are vividly illustrated in another Aztec codex, known as the Codex Mendoza, created in the 1540s.[35]

The indigenous people of Brazil, the Andes and Mexico created extraordinary artefacts from feathers. The Aztecs made feather cloaks, known as *tilmàtli*, that signalled the wearer's social status. The Tupi of Brazil constructed bonnets from parrot-down feathers (that were thought to symbolize a down-covered young parrot), as well as elaborate full-length body capes of Scarlet Ibis feathers that represented the 'sleek, structured profile' of the adult bird. Recognizing both the talent of the indigenous peoples and the meaning feathers held for them, the Spanish missionaries, in what was either a sympathetic or cynically motivated process known as 'accommodation', allowed local people to incorporate feathers into Catholic ceremonies, and to create featherwork versions of sacred images, including those of Jesus and the Virgin Mary.[36]

For the indigenous people of the Americas feathers were objects of worship, 'radiant and iridescent indices of the spirit world; ritual paraphernalia of sacred power; reifications of light, vision, and flight; embodied and ornate festive attire for elites; votive offerings; luxury material akin to precious metals, shells, and coloured stones'.[37]

The feathers used by indigenous peoples came from birds both dead and alive, and their plucked feathers 'were carried from the rainforest over the Andes as tribute to the Inca rulers, to be converted into artefacts in feather workshops in the highlands and along the Pacific coast'. An account from 1535 by Pedro Sancho de la Hoz, Francisco Pizzaro's secretary, mentions a warehouse near the Inca capital, Cuzco, in the Peruvian Andes, containing more than 100,000 dried birds whose plumage was to be used for making artefacts.[38]

As Columbus was about to set off on his transatlantic journey in August 1492, Queen Isabella I of Spain commanded him to bring back some birds for her. As luck would have it, on setting foot on Hispaniola he saw local people with tame parrots. Columbus and his crew were enthralled. Colourful, friendly and able to speak, parrots quickly became the ultimate souvenir. Exchanged for European goods, parrots (along with monkeys) made their way back to Spain, and subsequently elsewhere in Europe. Given initially to royalty

Aztec Warriors (Eagle Warrior at the left and Jaguar Warrior at the right) brandishing a macuahuitl (a wooden club with sharp obsidian blades) (The Field Museum Library).

and the very rich, the trade was such that within just a few decades parrots became almost commonplace, sliding downwards through the cultural hierarchy from one social class to another.

The Spanish conquerors were shocked by the sight of indigenous women feeding young parrots using the mouth-to-beak regurgitation of chewed food, similar to the beak-to-beak process that parent parrots use. This intimate parrot–person relationship meant that, as the birds matured, they remained tame and lived freely in the house as part of the family. They were more than mere pets: these parrots served a purpose, since their feathers were plucked at regular intervals to provide the raw materials for headdresses, capes and other featherworks. In 1587 the colonist Gabriel Soares de Souza discovered that the indigenous people were able to alter the colour of their parrots' feathers – usually from green or red to yellow – by periodic plucking. The process, known as *tapiragem*, has excited speculation ever since, and the colour change is now thought to be the result of localized trauma in which the plucking disrupts the orderliness of the nano-keratin structure in the feather barbules. Sometimes the effect was thought to be enhanced by the application of various potions – including that from a poison dart frog into the follicle from which the feather had been plucked. A change in plumage colour could also be achieved by feeding parrots on the fat of the redtail catfish – hardly something they would normally have ingested.[39]

The colour of a bird's plumage is created either by pigments or by microscopic structures within the feathers. Melanins and carotenoids are the main pigments. Melanins – manufactured by the birds themselves – are responsible for browns and blacks, while carotenoids – obtained from the birds' diet – comprise the reds and yellows. Structural colours, created by the nano-structure of the keratin within the feathers, are blues – often iridescent blues. Mixing this structural colour with a carotenoid-based yellow creates the iridescent blues, turquoise and greens that are the quetzal's magic.[40]

For the Tupi, Nahua and many other indigenous groups, feathers helped to transform the wearer into a spiritual version of the bird itself. Feathers signalled divinity and power, and were used in funerary contexts, in post-battle rituals and in sacrifices. As a remnant of those ancient traditions, Bororo children today in the Mato Grosso region of Brazil have toucan-down feathers stuck to them in a rebirthing ritual. The Bororo are known for their rich ceremonial life, and during such events adults wear feathered hats called *cocares* that hang from the top of the head to the shoulders, or sometimes almost to ground level. Most are made from parrot or macaw feathers, but those that are made from the feathers of the rare Harpy Eagle and reach to the ground confer the highest status.[41]

The close relationships of Amerindians with birds were interwoven with their attitudes towards people. In inter-tribal warfare, male prisoners were killed and consumed, while women and children were adopted and incorporated into the victors' tribe. Cannibalism allowed the consumers to assimilate some of their victims' strength. In much the same way, by wearing the pelt of a jaguar, or the plumage of birds such as eagles, shamans could assume the attributes of those species. By placing themselves inside animal skins, they felt that they had transformed themselves into that species: a jaguar pelt provided ferocity and strength; eagle feathers imbued the wearer with the bird's mighty essence; the glittering iridescence of the quetzal's feathers provided 'beauty, preciousness, fertility and transcendence'.[42]

Anthropologists in the past assumed that featherworks were usually worn by high-ranking men, but a remarkable account from 1557 of a ceremony leading up to the execution of an enemy captive, recounted by the Jesuit Antonio Blázquez, shows that this was not the case.

> Six nude women came by the public square . . . making such gestures and shaking movements that they really did seem like demons. From head to feet they were covered with red feathers. On their heads they wore *carochas* [Inquisition-like caps] of yellow feathers. On their backs they wore an armful of feathers that appeared like a horse's mane, and to animate the celebration they played flutes made from the shinbones of their slain enemies. With this attire they walked around barking like dogs and faking speech with so many mimes that I do not know with what I could compare them. All of these acts took place six or eight days before the killing.[43]

Accounts like these, featuring nudity, human sacrifice and cannibalism, fired the European imagination, and helped to create a market for travellers' tales as well as for featherworks themselves. When the Portuguese explorer Pedro Álvares Cabral turned up on the coast of Brazil in April 1500, a friendly encounter with the indigenous Tupi saw his men exchanging their headgear for featherwork hats. Soon after Hernán Cortés arrived in Mexico in 1519, when relations were still 'friendly', the Aztec emperor Moctezuma II gave him several exquisite featherwork crowns that he sent back as gifts for Charles V, who – impressed by their beauty – in turn gave them as gifts to other sovereigns.[44]

The appearance of these fabulous featherworks in Europe signalled the extraordinary success of the Jesuit enterprise. Regarded with amazement, featherworks were eagerly acquired by the wealthy for their then-fashionable cabinets of curiosity. Some also formed the basis for contemporary ceremonies, including the barely believable 'Queen of America' Mardi Gras procession that took place in the German Duchy of Württemberg in 1599.

This was a pageant designed to consolidate Duke Friedrich's authority following a political dispute, in which Friedrich and his courtiers dressed in Tupi feather costumes and paraded in front of a crowd of some 6,000 awestruck spectators, including dignitaries from all over Europe. Awestruck they should have been, for the Queen of America, borne aloft on a ceremonial float, was none other than Friedrich himself in the guise of a voluptuous, naked, native woman. The transformation – an act of great symbolism – was achieved by Friedrich sporting a full body suit of leather, complete with papier mâché breasts and adorned with feathers, tattoos and gold ornaments.

I wonder how much of this complex symbolism was deliberate and how much of it was understood by those who witnessed it. Nudity was sometimes construed as childlike and innocent, reflecting the Christian idea of existence before the Fall, but in this case Friedrich's (apparent) nakedness was designed to reflect the savagery and depravity of indigenous peoples. The flouting of priceless featherworks in the procession was not only a statement of Friedrich's wealth, but also, by his appearing as a woman, a quest for male adulation. In addition, it symbolized his superiority over his subjects at home, as well as the indigenous American peoples – Europe's bride – ready to be taken.[45]

That whole bizarre business was choreographed using a series of miniature watercolours 'depicting the sequence of characters in Friedrich's procession, each labelled with the name of the person who wore the costume. The participants were ordered in a clear hierarchy of social precedence, thus re-enacting the political structure of the court.'[46]

The 'Queen of America' pageant anticipates the crass cultural appropriation of the French aristocracy – Marie Antoinette and her Versailles entourage – as they frolicked about dressed as peasant milkmaids and shepherdesses, while the peasants in pre-revolution France were starving. It also provides an example of the appropriation of indigenous artefacts into a completely different culture, with an almost completely different meaning. The only overlap between the two is perhaps the way in which feathers reflected wealth and power. The importation of New World featherworks into Europe was a journey of dilution, degradation and eventual loss of meaning – a banal Disneyfication of indigenous dignity.

The conquest of the Americas was devastating for the indigenous peoples. As the Spanish deliberately set out to destroy local cultures, their colonial appropriation was accelerated by the unintentional introduction of European diseases to which indigenous peoples had no immunity, resulting in a massive decline in their populations. And the reason for all this was greed: the acquisition of resources – gold and silver initially, later timber, sugar, cacao, all produced by the labour of slaves, either local or imported from Africa. The invading powers appropriated another resource: local knowledge, some of which at least, such as that relating to medicinal plants, was of huge commercial potential, for example tobacco, and guaiacum as a cure for syphilis.[47] Other knowledge, including the description and identification of new species of birds, might seem to have had 'only' scientific value, but this too was profitable through the sale of books, including those of Nieremberg, Piso and Marcgrave, and Ray's *Ornithology*.

European naturalists were, often unwittingly, part of this colonial exploitation through the appropriation of indigenous knowledge. As well as in the Americas, this was also true elsewhere, including India (as we will see in Chapter 9) and Africa. In recent decades, this arrogation of local knowledge by colonial powers has been criticized for 'purposely effacing the lives of subaltern actors' and because 'The role of indigenous informants all over the world in these narratives, when acknowledged, is often reduced to that of victims: forced mediators between the objects of nature and their study by European men of science.'[48] In a detailed assessment of the role of indigenous informants and assistants in the development of early twentieth-century African ornithology, Nancy Jacobs comments on how 'assistants were subordinates even while they were experts', referring to the 'false innocence' of natural history writing – compared with the imperial expansion and enslavement.[49]

A major criticism has been that European naturalists filtered out much or all of the original meaning from indigenous knowledge. The use of feathers from different bird species and their juxtaposition within a headdress, for example, had – judging from the information recorded in various codices – specific meanings that the transcribers rarely bothered to record. Historian Marcy Norton illustrates the problem with the example of a European adventurer:

> John Ogilvie . . . who lived among [the] Waiwai in British Guiana in the early twentieth century, complained that in order to purchase a feather headdress, he was 'subjected [to] the history of the hat itself! I was taken on a verbal hunt

after each bird . . . how and where it was shot, and countless long-winded details.'[50]

I have some sympathy with Ogilvie; one's upbringing and, in my case, scientific training, work against embracing and really appreciating such information. Having studied birds in the Canadian Arctic and worked with local people there, I always wanted to understand the bird folklore of the Inuit, but their stories seemed to lack the structure and coherence that we have been trained to think of as 'meaningful', and so to us those stories can seem pointless, even though to the orator they are not. Despite his apparent impatience, Ogilvie, was well versed in Waiwai culture, spoke the language and had a deep respect for the people, so he knew full well what his informant was doing. Shape-shifting was what occurred when Indians wore certain feathers, and shape-shifting is what is needed to understand the way different cultures see and value birds, even though that may be difficult.[51]

To be fair to Ray and Willughby, in the preface to the *Ornithology*, they are very clear that their objective was scientific: 'We have wholly omitted what we find in other authors concerning . . . hieroglyphics, emblems, morals, fables, presages, or ought [*sic*] else appertaining to divinity, ethics, grammar or any sort of humane learning.'[52]

While the spiritual meaning of birds has been filtered out from most European accounts of indigenous natural histories, so too in many cases has the essential debt owed to indigenous informants.

The appropriation of indigenous knowledge by colonial powers has been likened to the taking of unique physical artefacts, such as the paintings hacked from the walls of Egyptian tombs (Chapter 2), but also non-unique anatomical specimens such as bird skins. There's a difference in that, in principle at least, artefacts can be returned, while knowledge cannot. Certainly, the Europeans benefited from the local natural-history knowledge they acquired, but the main issue seems to be the lack of credit of the source of this knowledge.[53] Although this is often absent, it is also true that individuals like Bernardino de Sahagún and Francisco Hernández in Mexico in the 1500s, and the ornithologists François Levaillant and Con Benson in Africa in the late 1700s and early twentieth century, respectively, were explicit and generous in acknowledging their debt to indigenous assistants and informants.[54] These ornithologists knew they were standing on the shoulders of indigenous giants. Given the cascade of knowledge through successive publications that characterizes both the study of Mexican birds and current science as a whole, perhaps we should not be too critical of John Ray for not crediting indigenous knowledge in his account of the Resplendent Quetzal.

Few researchers in either the sciences or humanities are entirely free from bias. Notwithstanding any possible antipathy he had towards Spanish researchers, John Ray changed the way people in Europe and elsewhere related to birds. After Willughby's death, Ray spent much of the rest of his life loyally and assiduously writing up and publishing the work he and his colleague had begun: birds in 1676, fishes in 1686 and insects in 1710 (published after Ray's own death in 1705).[55] And in between these monumental zoology projects, Ray continued to squeeze in works about plants and, most famously of all, about God.

The Wisdom of God Manifested in the Works of the Creation, published in 1691, had its beginnings in lectures Ray gave in Cambridge in the 1650s, but it was fertilized by, and grew from, the work he and Willughby conducted together. *The Wisdom of God* switched focus from identification, description and classification, as in the *Ornithology*, to interpretation. Essentially, to features that fitted birds – and other animals – to the environment in which they lived: adaptations. *The Wisdom of God* changed both the course of natural history and people's attitudes to it.

Deeply religious and profoundly inspired by the natural world, Ray found a clever way to combine these two world views, building on the work of several others that celebrated the beauty and wonder of nature – of God's world.[56] It was also an empathetic view, vigorously rejecting the Catholic Descartes's notion of animals as soulless 'automata' – machines devoid of feeling. This new way of looking at the natural world was referred to as 'physico-theology' or, more commonly, 'natural theology'. Its focus was on adaptations, like the woodpecker's long tongue, the webbed feet of waterfowl and the cryptic colouration of nightjars. These kinds of attributes were evidence of God's providence in creating animals perfectly suited to their environment. The woodpecker's tongue allowed it to be 'thrust deep into the holes of trees to draw out' insect larvae; the webbed feet of ducks and other water birds facilitated their swimming and nightjars' perfect camouflage allowed them to sleep on the ground during the day unnoticed by predators.[57]

Like many of his predecessors and contemporaries, Ray recognized the utilitarian value of birds, but he added a new twist: animals were not designed by God explicitly for man's use, but rather, they were 'by the wit of man accommodated to those uses'. In other words, with little or no help from God, man had worked out that:

> Birds are of great Use to us; their Flesh affording us a good Part of our Food, and that the most delicate too, and their other Parts Physick [medicine], not excepting their very Excrements. Their Feathers serve to stuff our Beds and Pillows, yielding us soft and warm Lodging, which is no small Convenience and Comfort to us, especially in these Northern Parts of the World. Some of them have also been always employed by Military Men in Plumes, to adorn their Crests, and render them formidable to their Enemies. Their Wings and Quills are made use of for Writing-Pens, and to brush and cleanse our Rooms, and their Furniture.[58]

While Ray does not explicitly mention falconry in his 'value of birds' list, perhaps it is implicit in 'Exercise, Diversion, and Recreation'. But Ray does clearly recognize the symbolic or decorative uses of birds 'employed by Military Men' – which might be a nod at least towards the use of quetzal feathers by Amerindians. He ends, however, with a strong case for valuing the sheer beauty of birds through their song and appearance:

> Besides, by their melodious Accents they gratify our Ears; by their beautiful Shapes and Colours, they delight our Eyes, being very ornamental to the World, and rendring the Country where the Hedges and Woods are full of them, very pleasant and chearly, which without them would be no less lonely and melancholy. Not to mention the Exercise, Diversion, and Recreation, which some of them give us.[59]

Ray's wide-ranging *Wisdom of God* is nothing less than a history of science, but it also marked a new and 'vastly important' shift in attitude to the natural world: 'The delight in the worth of the [natural] world as aesthetically satisfying, intellectually educative and spiritually significant reflects the best Hebrew, Greek and Christian thought, but it is in strong and striking contrast to the philosophy and religion of both the Catholic and Protestant tradition.' Prior to this the natural world was irrelevant or even hostile to religion: its beauty was 'a temptation and its study a waste of time'. Ray's *Wisdom of God* gave the development of science a status and sanction of the highest value; and, so far as Britain was concerned, freed it from the conflict with ecclesiastical authority.[60]

By 'drawing attention to the unity of nature, to the problems of form and function, to adaptation, and to a great number of strange . . . phenomena', Ray gave the 'first strong impulse to the scientific movement of today'.[61] He also inspired subsequent generations of naturalists to 'leave their libraries and go into the field to carry out studies on . . . living birds'.[62] A tiny handful of individuals did just this, breaking out of the ornithological straightjacket to look at live birds either in their natural habitats or in captivity, or, in the case of the extraordinary Baron Ferdinand Adam von Pernau, in both. Pernau, whose writings were published anonymously in the early 1700s, was a careful and perceptive observer and among the first to speculate sensibly about how birds acquire their songs. Another pioneer of bird behaviour, Nicolas Venette – a retired medic – made observations in the 1690s of his captive Common Nightingales identifying the fundamental phenomenon of migratory restlessness, and proffered eminently sensible suggestions to explain it.

Then there was Luigi Marsili, an Italian aristocrat, soldier and scholar, who in the early 1700s undertook the first field study of birds' nests and eggs. In 1742–3 the German natural-theology enthusiast Johann Zorn produced a two-volume book, *Petino-Theologie* (*Winged Religion*), that included hundreds of anecdotal observations of wild birds, including the idea that the young Common Cuckoo mimics an entire brood of warbler chicks to dupe its foster parents into even greater efforts – later confirmed by my colleague from PhD days, Nick Davies. To see how the observation and study of live birds in their natural habitats fostered a distinctly empathetic relationship with birds, we will have to wait until Chapter 11.

John Ray's *Wisdom of God* was a bestseller, and for years afterwards he was kept busy producing new editions. The book marked the beginning of field biology – watching birds and other animals in their natural environment. It inspired others to do the same, including the revered Reverend Gilbert White. Ray's *Wisdom* also inspired William Paley, who a century after it was first published plagiarized it in his equally successful *Natural Theology* of 1802. This in turn was read by Charles Darwin as an undergraduate at Cambridge, who – not without controversy – went on to turn the idea of natural theology on its head by identifying natural selection rather than God as the agent of adaptation.

7. Depending on Birds: Inconspicuous Consumption

> Passing along wee saw some round hills of stone, like to grasse cockes, which at the first I tooke to be the worke of some Christian . . . I turned off the uppermost stone, and found them hollow within and full of fowles hanged by their neckes.
>
> Abacuk Pricket (1610)[1]

The Ornithology of Francis Willughby was written in Latin, the scientific language of the day. But things were changing, and soon after its publication, Ray's friend Martin Lister urged him to produce an English edition. Thank goodness he did, for without it, Willughby and Ray's ground-shifting contribution to ornithology might have gone unnoticed. Ray used the opportunity to rectify a few errors, but in the hope of boosting sales he also made some topical additions, including a recent account of the seabirds from the faraway Faroe Islands written by a Danish priest.

Lucas Debes arrived on the Faroes in 1651, where he was to live almost continuously for the next twenty-four years. A gentle, inquisitive and well-educated man, his compelling account of the Faroese people reported how 'God hath plentifully blessed the Land with several sorts of Fowle, whereof the greatest part serveth for the food of man'.[2]

Few cultures have depended so much on birds for sustenance as that on the Faroes. This archipelago of thirteen main islands lies in the storm-battered seas equidistant between Scotland, Norway and Iceland. The barren, mountainous land supports an impoverished fauna and flora, but the surrounding seas were once among the most biologically productive marine areas anywhere on earth. A massive spring bloom of plankton fuelled the fish populations that in turn nurtured vast numbers of top predators, including whales and almost unimaginable numbers of seabirds.

Offshore islands are the perfect breeding place for seabirds. Devoid of terrestrial predators like foxes, weasels and rats, the birds – and, most important, their eggs and chicks – are relatively safe, and the islands are completely surrounded by potential feeding areas. The Faroes' stupendous cliffs and high grassy slopes provide a superabundance of breeding sites. As their name implies, seabirds are tied to the sea for their food and have evolved to make the most of this feeding niche. Think of albatrosses and Northern Fulmars, whose long, stiff wings allow them to glide almost effortlessly over tremendous tracts of ocean to sniff out their prey. Or the Atlantic Puffins, Razorbills and Common Guillemots whose dumpy little wings won't carry them quite as far, but serve admirably as paddles, enabling them to dive to extraordinary depths in search of fish and plankton.

At the height of the breeding season a seabird colony is like a modern-day city: busy, smelly and noisy. Thousands of birds swirl above and around the breeding areas; many more are sitting on the sea, or on the tidal rocks at the base of the cliffs; a chorus of calls that range from the Northern Gannet's guttural and urgent *hurra-hurra*, to the Black-legged Kittiwake's melodic falsetto *kittywake* slicing through the noisy hubbub. There's also the ammonia-rich aroma of seabird shit (which I love, by the way) covering the ledges,

The Faroes, looking west from Vagar, with Tindhólmur (and Eagle's Peak) in the mid-foreground and Mykines on the right distant horizon (photo: Tim Birkhead).

splattered and splashed in endlessly artistic designs, dribbling eventually into the sea, where it provides the nutrients that feed the phytoplankton that drive this entire, wonderful ecosystem.[3] In winter, however, with the birds gone, across the oceans, a seabird colony can seem an eerily desolate place.

No one knows for sure when people first arrived on the Faroe Islands. New archaeological discoveries keep pushing the date back, but it was sometime in the first millennium AD, long after El Tajo's Neolithic people had disappeared. Some Faroese archaeological artefacts date back to between the fourth and sixth centuries AD, and the first colonists are thought to have been Irish monks like Brendan, who in his sixth-century account refers to what may have been the Faroes as 'the island of sheep' and 'a paradise of birds'. His description could have referred to almost any of the North Atlantic islands that until recently were all a 'paradise of birds', whereas today many of these seabird colonies, especially those on the Faroes, have declined so much that even in the height of the breeding season they exude an air of winter desolation.[4]

That sense of despondency is exaggerated by the sheer size of the cliffs, whose now tiny populations of birds can barely be heard or seen from the cliff tops. In fact, my overwhelming feeling when I visited the Faroes in May 2019 was that one could be there and barely know – based on what you can see, hear or smell – that these are seabird islands.

Yet, talking to local people, and especially those on the more remote islands, it becomes clear that fowling – the catching and eating of seabirds – still lies at the heart of Faroese culture. Descending the vast cliffs, risking life and limb to take eggs and adult birds, imposed a way of life very different from that of Continental farming communities. Fowling required teamwork, and while it was invariably men who scaled the cliffs, women often helped to haul them back up again. Climbing was both physically demanding and risky, providing both status and a deep sense of camaraderie among those involved, reinforcing a peculiarly distinctive, close-bonded way of life.

Lucas Debes's account is the first of Faroese fowling, seabird biology and culture. The only known portrait of him was lost in about 1870, but a relative who saw it described him as a tall man with a black beard and long black hair. His *Description of the Islands & Inhabitants of Foeroe* published in 1676, is a tour de force: perceptive, objective and written with extraordinary enthusiasm and empathy for ordinary Faroese folk:

> Eatable Sea Fowles are found in great quantities . . . the Skrabe [Manx Shearwater], Lunde [Atlantic Puffin], Lomvifve [Common Guillemot] . . . the aforesaid sorts of birds lay . . . but one egg, and get but one young every year, and though they be those that chiefly are sought for, and there be well taken of them a hundred thousand every year . . . by the admirable providence of God, they are so plentiful that

> they in clear weather can darken the shining of the Sun as it were with a thick Cloud, making such a terrible noise and sound with their wings in flying, that they, who hear it, and do not know the cause thereof, would not think otherwise, but that it were thunder.[5]

His account includes a lovely piece on the Great Auk:

> Here cometh also a rare water Fowle, called Garfugel [the Great Auk], but it is seldom found on Clifts under the promontories, it hath little wings and cannot fly; it stands upright and goeth like a man, being all over of a shining black colour, except under the belly where it is white; it hath a pretty long raised Beak though thin toward the sides, having on both sides of its head over the eyes a white round spot as big as a half Crown, showing like a pair of Spectacles . . . I have had that Bird several times, it is easie to be made tame, but cannot live long on Land.[6]

Debes was twenty-nine when he arrived on the Faroes in the summer of 1651 to serve as curate to the resident vicar. It was only then that he discovered that his predecessor had died some months earlier and that he should marry his widow – as was the Faroese tradition – and become the vicar of Tórshavn, the Faroes' capital. Instant promotion, accompanied by immediate responsibility, since his new wife had nine children, for whom he now had to provide. As a result, money was tight and Debes was often in debt.

Energetic and smart, Debes quickly came to like and respect the Faroese people. He was appalled by their treatment at the hands of their Danish landlord, Kristoffer Gabel, and took it upon himself to defend them. The period of the Gabel family's fiefdom between 1655 and 1673 – coinciding with Debes's tenure – was said to be the bleakest time in the Faroes' history. Gabel never once left Denmark to see how his tenants managed to survive under his oppressive and greedy rule. Debes had a long battle with Gabel's bailiffs – one of whom raped Debes's step-daughter.[7]

Part of his motivation for writing his *Description of the Islands & Inhabitants of Foeroe* was to publicize the plight of the Faroese people, but the single most remarkable aspect of Debes's account is its insight into their overwhelming dependence on seabirds. His descriptions of how the local people caught and used birds makes him one of the first cultural anthropologists, yet his name is rarely mentioned in such circles. Written during the 1660s, Debes's book was first published in Danish in 1673 and, unexpectedly perhaps, translated into English in 1676.

The way this came about is one of the sparkling threads in the emerging tapestry of the Scientific Revolution during the 1600s. Soon after the Royal Society of London had been formed in 1660, its dynamic and enthusiastic secretary Henry Oldenburg launched the world's first scientific – and still active – journal, the *Philosophical Transactions*, as a way of transmitting and preserving details of new scientific discoveries. Oldenburg also helped the society's Fellows to make new discoveries by identifying topics on which more needed to be known. To do this he devised sets of structured queries for the Fellows to seek answers to during their travels both at home and abroad. The topics of these 'queries into natural history' were broad and included things like life in different countries, the effects of cold temperatures, activities such as mining, or sometimes simply requests for specimens, such as 'spiders that are said to spin webs so strong, as to catch birds'.

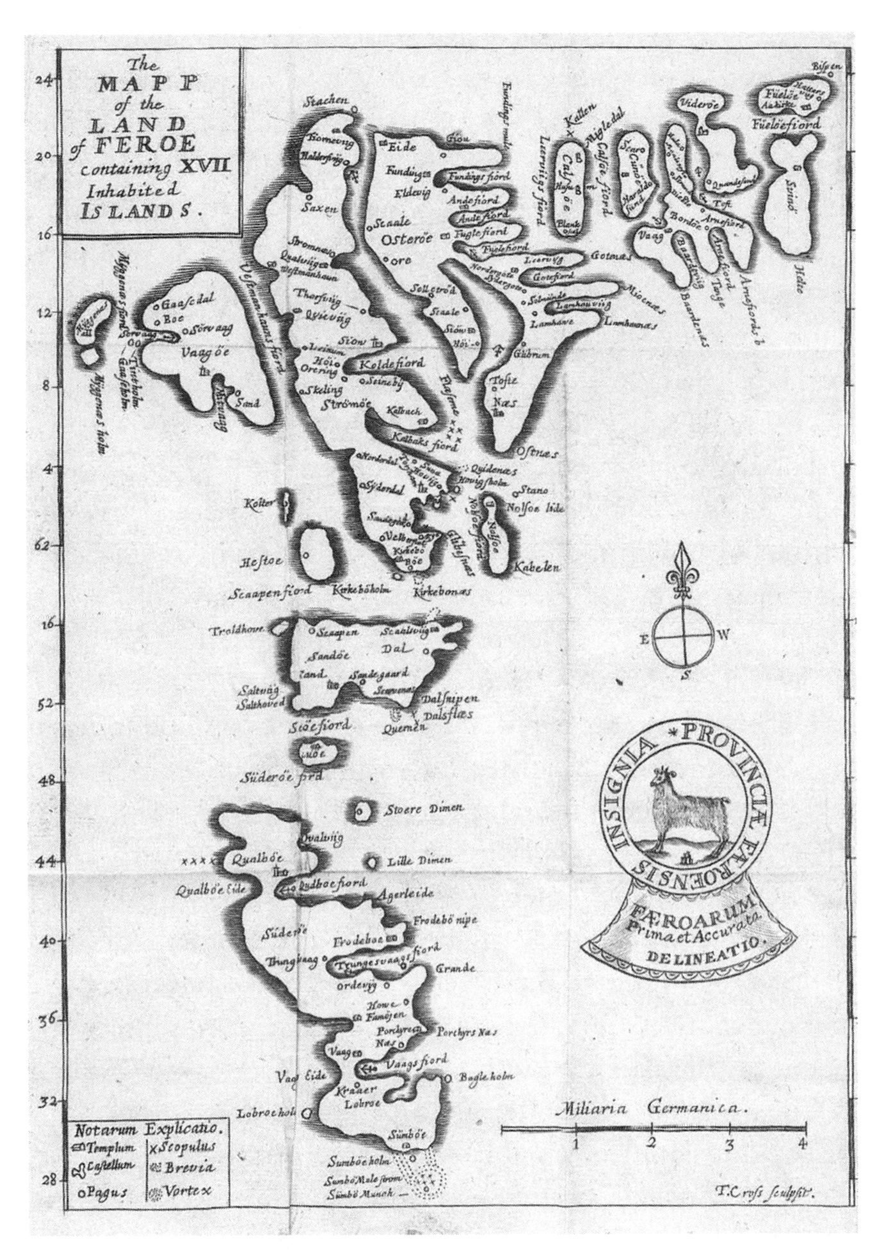

Lucas Debes's map of the Faroe Islands (from Debes, 1676).

Oldenburg was interested in Denmark and its natural history, and in 1672 his colleague Thomas Henshaw, a Fellow of the Royal Society, and royal envoy to Denmark, based in Copenhagen, wrote to Oldenburg to tell him that he had met someone he called 'Monsr Gabelli' (Kristoffer Gabel), 'proprietor of ye island of Fero'. Recognizing a rare opportunity, Oldenburg asked Henshaw if he would create a set of queries relating to the Faroes, about which nothing was then known. He did so, but much to his annoyance no one, including Gabel, was interested. Then, by chance, one of Henshaw's Danish colleagues, Rasmus Bartholin – physician and professor at the University of Copenhagen – told him about a Danish priest living on 'ye island of Fero [who] had recently written a book in the Danish Toung concerning that island'. Henshaw contacted Debes – they may even have met when Debes was in Copenhagen – who subsequently provided answers to Henshaw's queries.

Delighted by this result, Henshaw wrote to Oldenburg in 1673, referring to 'Magister Lucas Jacobi Debes . . . a man ingenious and curious' who had written a large book about the Faroes, 'but ye pity is, it is written only in Danish'. Henshaw wondered whether by giving Debes 'some present I can prevale with him to make some abridgement of it in Latin or high Dutch'. But no abridgement was necessary, for in the meantime Oldenburg had arranged for John Sterpin – a Scot born in France but educated in England and resident in Copenhagen – to undertake an English translation.[8]

The translation was published in 1676 and a review of it in the Royal Society's journal commented that the Faroes is a country 'well stored both with Land- and Sea-fowl', and I suspect it was this that caught the attention of the eminent English physician, naturalist and member of the Royal

Society Martin Lister. With great excitement, Lister wrote to tell John Ray about the book, suggesting that he include some of it in the English version of the *Ornithology*.

On reading Debes's book, Ray immediately identified with the author's scholarship and the novel information on Faroese seabirds it contained, and copied a section of the book, verbatim – albeit, with appropriate credit – to add to the new edition of the *Ornithology*.

It is little wonder that Debes wrote so enthusiastically about Faroese birdlife, for during his education in Copenhagen he had been tutored by the renowned naturalist Ole Worm, whose main claim to fame, subsequently at least, was his huge cabinet of curiosities. Among much else, Worm kept a Great Auk as a pet, sent to him by Debes, noting with amazement how the bird could swallow a herring whole.[9]

The Faroese culture was a deeply religious one, and there was no fowling on Sundays, yet they were permitted to hunt pilot whales on the Sabbath (the difference, presumably, being that the birds were tied to their colonies and would still be there on a Monday, whereas there was no guarantee that the pilot whales, once seen, would still be around the next day). And, of course, the pilot-whale hunt, or 'grind' – already well established in Debes's day – had the potential to provide a mountain of meat.[10]

Sixteen different seabirds breed on the Faroes and their affinity with the Faroese people is exquisitely captured in four fragile nineteenth-century watercolours by the self-taught artist Didrik Sørensen. His paintings, now in the Faroes National Gallery in Tórshavn, show the birds he and everyone else would have been familiar with: the Common Guillemot, Atlantic Puffin, Razorbill, Eurasian Oystercatcher, Bonxie (Great Skua), European Golden Plover and

Ole Worm's Great Auk (from his account of 1655) was a gift sent from the Faroes to Denmark by his pupil Lucas Debes. The white ring around the bird's neck was probably a collar by which Worm controlled it, but subsequent authors erroneously assumed it to be part of the bird's plumage.

the Faroese speciality, the pied Northern Raven. Painted in profile on beige paper (or possibly, paper that was once white, but has now assumed an antique hue), Sørensen's birds are naive, but enchanting.[11]

Until very recently, the nutritional foundation of Faroes society rested largely on the wings of three seabirds: the Atlantic Puffin, Common Guillemot and Northern Fulmar.

To some non-Faroese people, the idea of eating a puffin is as repulsive as suggesting we eat a Labradoodle. But for tourists visiting Iceland novelty routinely overrules disgust, and this once-traditional dish is popular with visitors. On the Faroes, too, the puffin has been an important part of the diet, and probably for the last thousand years – not to titillate the tourist palate, but through necessity. An English visitor who lived with a local Faroese family in the 1940s considered puffin served with boiled potatoes, thick brown gravy and rhubarb to be 'delicious – rich and tender, and very tasty', and preferred by the Faroese to either the guillemot or Razorbill.[12] And I, too, have eaten puffin – in the 1970s, after being abandoned on a seabird island off the coast of Labrador by fishermen who were supposed to collect us. Having eaten all our food we ate some puffins – to me they tasted like a liver-flavoured steak.

The puffin has been the most important of the seabird species caught and eaten on the Faroes and is much more readily and safely harvested than either the cliff-breeding guillemot or fulmar. Puffins lay their eggs in burrows on accessible grassy slopes, and can be hauled or, more brutally, hooked from their burrows, or caught as they fly over the colony using the long-handled *fleyg* net – two techniques that have been in use for at least the past three centuries and probably much longer. Debes describes how the Faroese caught and processed puffins:

> The men take them flying, which is done in this manner: they have a long Pole at the end of which there is a hoop drawn

> over with a net, whereof the meshes are almost as big as the quarrels of a Glass Window, being like the Net, wherewith they take Shrimps in some places; and this they call a Stang or Staffe; with this Staffe the Fowler sitteth on the Clift, or in the Ures between the great Stones, where he knoweth most Fowl to come, which they call flight places; and when the Lunde [puffins] cometh flying either from or to the Land, he lifts up the Staff and the Net against the Fowle, and when he hath got it into the Net, he turneth the Staffe about, that it may intangle it self the better therein; a man being sometimes able in that manner to take 200 Lundes in very short time, besides those that are taken in their Nests.[13]

The harvest in the 1600s was huge, but unknown outside the Faroes. In the early 1900s (and probably for many years before) somewhere between 400,000 and 500,000 puffins were taken each year. To sustain this annual toll, the Faroes puffin population must have comprised around two million pairs. But today it is much smaller. Over the past 120 years the puffin harvest has declined – to 154,000 in 1940, 95,000 in 2000 and a mere 10,000 in 2007. To a large extent this change is the result of a decline in the puffin population itself, attributable to at least two factors (other than the harvest): rats and a lack of food.[14]

Brown rats first arrived on the Faroes in 1768. On those islands infested with rats, the ground-nesting seabirds – puffins, Manx Shearwaters and storm petrels – all disappeared. Rats eat seabird eggs and young – an all-too-familiar story across the world.[15]

Just as devastating as the rats on land have been changes at sea – shifts in the distribution and abundance of the fish that puffins rely upon to breed, caused by climate change and

Part of the Danish garrison commander C. V. L. Born's map of the Faroes from 1791, accompanied by vignettes of local life: here, the different ways of capturing seabirds (from Jensen, 2010).

overfishing. The lack of food for breeding puffins on the Faroes was first noticed in the early 1970s, when breeding success was much lower than normal. In subsequent years the puffin breeding season was either delayed or the birds failed to breed at all – as in 2007. In other years, the chicks from the entire Faroes population starved to death. The same disastrous effects have been recorded in other North

Atlantic colonies, including those in Norway, Iceland and Scotland, and are part of climate-change shift in the abundance and distribution of plankton and fish.[16]

After centuries of puffin hunting with no detectable decline in numbers, the Faroese blame the rats and the oceanographic changes for the decline in puffins. Indeed, their method of harvesting puffins – the *fleyg* – deliberately targeted non-breeding birds. Puffins do not breed until they are six or seven years old, but from the age of three or four they start visiting the colony, flying round and round in huge 'wheels' over the breeding areas. By mainly taking these immature birds, and avoiding birds with chicks, the Faroese have tried to safeguard the breeding stock.

Something positive has to have come out of killing and consuming all those puffins, and in recent years it has. As anyone who has spent time at a large puffin colony during the breeding season may know, sometimes, among the immaculate masses, there stands the occasional 'dark-faced' puffin. These are birds that have dark grey (instead of white) feathers on the face and lack the fleshy ornaments above and below the eye and the yellow-edged, blue-grey plates at the base of the beak. Remarkably – why did no one think to check this before? – those pale yellow borders to the bill have recently been shown to glow in the dark and may help the chick in its burrow find its parent's bill at feeding time.[17] The standard and self-evident explanation for these distinctive dark-faced birds is that they have failed to assume their breeding bling. But, as a scientist, one knows that this is a rather wishy-washy explanation. However, by dissecting no fewer than eighteen dark-faced birds killed by Faroese fowlers over the years, local ornithologist Jens-Kjeld Jensen has made a remarkable discovery. What he found was that all the dark-faced puffins are old (judged by the number of grooves on the red part of the bill), and they are all females. What is going on? The answer is that, as female birds age, their ovaries sometimes cease to function properly – and indeed, the ovaries and oviducts in every one of these dark-faced puffins were in a bad way. Non-functional ovaries are unable to produce oestrogen, and in birds such as pheasants and peafowl it is known that the lack of this hormone causes females to change sex and moult into male plumage – much to the bewilderment and consternation of their owners. This occurs because male plumage is generated by a *lack* of oestrogen, suggesting that in puffins the moulting and replacement of those gaudy bill plates is also determined by this particular hormone.[18]

It is mid-afternoon as the boat pulls into the tiny island of Skúvoy with its huddle of colourful houses. Walking up the steep track towards the furthest house, I am amazed by the sight of another huddle – of Faroese gnomes. Are these, I asked myself, the hidden people of Faroese folklore – the Huldufólk? The plastic gnomes are in the garden of Inga's house, where I am to stay. She greets us warmly, and after seeing our rooms, my friends and I head off in the bright afternoon sun to the seabird cliffs, dodging the aggressive Bonxies on the way. Breakfast the next morning is a spread of duck eggs, Inga's own *skerpikjøt* – air-dried mutton, created in a similar way to Serrano ham, but more of an acquired taste – toast and orange juice. Reaching into her fridge, Inga proudly pulls out a tray of fulmar eggs, her share from the very recent harvest, that she clearly has no intention of sharing with her bed-and-breakfast guests. On our telling her that the previous afternoon we had walked to the seabird cliff known as Høvdin at the north end of the island, where guillemots and kittiwakes breed, Inga responds with a deep sigh. 'It makes me want to cry to see so few guillemots there now,' she says. Getting up, she moves towards the kitchen window and looks out over the sea: 'From here,' she says, 'I used to see huge flocks of guillemots . . . but not any more.'

It is true, the past seventy years has seen a massive decline in Common Guillemots on the Faroes. Estimates of numbers are crude because this is a difficult place to conduct a census, but the decline is obvious. So obvious, in fact, that most forms of guillemot harvesting – the taking of eggs, the shooting of adult birds at the colonies, and the use of noose-boards to harvest adult guillemots at sea – have all ended. What has not stopped is the shooting of guillemots at sea during the winter months.

The best guess for the numbers of guillemots breeding on Skúvoy – the main guillemot island – prior to 1950 was two million birds; in 1961 there were 500,000, a decade later, in 1972, about 214,000, and by the mid-1980s numbers had fallen to 75,000, where they remained until the early 2000s, when the birds were last surveyed.[19]

The guillemots shot at sea during the October to January hunting season are Faroese, but these also include birds from Norway, Iceland, Greenland and Britain. The number killed is unregulated and unknown. I know from geolocation tracking that some of 'my' guillemots from Skomer Island come to Faroese waters in winter.

Inga brings out her old photograph albums from the 1960s to show us herself as a six-year-old on the cliff top at

Høvdin, with family and friends arranging the mass of recently collected guillemot eggs like so many toy soldiers in upright, regimented blocks. At that time, some 20,000 guillemot eggs were collected each year on Skúvoy, and at another island it was 40–50,000, so the entire Faroese total must have been well over 100,000 eggs each year. Inga tells us that she rates boiled guillemot eggs as superior to both fulmar eggs and her own duck eggs. As for the adult guillemots, the best bit, she says, is the brain.[20]

Guillemot eggs are large and nutritious; the yolk comprises about one third of the egg, as in a chicken, but it is twice as big. The 'white' or albumen is translucent blue and, unlike a chicken or duck egg, never quite sets, even when 'hard' boiled. The same is true of other seabird eggs, fulmars and gulls, and obviously reflects a difference in the composition of the albumen,

Returning bird-catchers Oddmar Poulsen and Niclas Lyngve on the island of Hestur in the Faroes in 1956 (courtesy J.-K. Jensen).

but to what purpose, I wonder. Guillemot eggs are tough; thick-shelled – they have to be because they are laid and (when not 'harvested') incubated on bare rock – and this means they can be transported with little risk of damage. Many of the eggs from Skúvoy were sent to shops in Tórshavn where they were sold for a few krónur each (1 krónur = 10p); the rest were eaten locally, either immediately or later in the year.

A mixture of peat ash and seawater seals the pores in the eggshell and – remarkably – keeps guillemot eggs in an edible state for many months. On certain islands it was a tradition to eat some of the previous year's eggs as the new eggs were being harvested. No effort was made to test the freshness of the eggs before preserving them, and no one seemed to mind if the egg contained a small embryo before eating it.[21]

In the 1940s, even the guillemot's skinny little chicks were eaten. When they left the colony at just a quarter of adult weight and flightless, the young birds were herded into narrow inlets and killed with a stone attached to a fishing line: 'Sometimes more than a thousand are slaughtered in this manner in a day.'[22]

The Faroese guillemot decline started during the Second World War, when food was scarce. Supply ships were sunk, and the Faroese turned to what was available locally. Prior to this, the level of seabird harvesting was assumed to be sustainable. Remove a guillemot's egg and it will produce a replacement some two weeks later. The Faroese only ever made a single visit to any particular colony to take guillemot eggs, so it is unlikely that egg harvesting had a major impact on the population – unlike colonies elsewhere.[23] Catching adult guillemots – with the *fleyg* – at the colony was more destructive since, unlike the situation with puffins, it killed the breeding stock, but it was less destructive than shooting adult birds at the colony. Shooting really was the most short-sighted form of harvesting, for not only did it kill adult birds, but also each shot fired triggered a mass panic among the other birds, causing hundreds of them to lose their eggs. In Greenland, local people had for centuries harvested eggs and adult Brünnich's Guillemots more or less sustainably, but once they started to use shotguns to kill breeding birds the effect was catastrophic, with many colonies spiralling rapidly to extinction.[24]

On the Faroes it was realized that shooting at the colonies was like shooting yourself in the gonads, and so in the 1870s there was a ruling that 'no gun may be fired within a distance of two English miles direct to sea, and one mile on each side of a breeding colony'.[25] But there's more than one way to catch a bird.

Noticing that guillemots could not resist hopping out of the sea to stand on a piece of flotsam, special boards covered in horsehair nooses were constructed and set afloat near the breeding colonies. First used in the late 1700s, the technique was considered inefficient and soon abandoned. But when reintroduced in 1921, its effect was devastating. Coming back at the end of the day, there could be as many as twenty birds – most still alive – snared by a leg on each board.

The Faroese immediately saw the advantages of this new technology and, knowing the tides, simply cast as many as thirty boards per boat into the sounds between the islands. No one knows how many birds in total were taken in this way, but two brothers who shared a boat and kept detailed records of their catch over twenty-eight consecutive years snared 380,000 guillemots. It is little wonder the colonies declined.[26]

By the late 1960s (by which time horsehair nooses had

been replaced by monofilament line) it was clear that the colonies were in trouble, and in 1979 snaring was made illegal. Too little too late, perhaps.[27]

Closely engaged with the Faroes' guillemots was a Danish biologist, Arne Nørrevang, a regular visitor to the Faroes from the 1950s, who moved there permanently in 1983. He is someone I would like to have met, because without knowing it, he helped to shape my career. In 1958, while in his twenties, Nørrevang published a study on the biology and behaviour of the guillemot in the Faroes that included some particularly exciting observations. Fourteen years later, in 1972, my PhD supervisors had already decided that the topic of my research would be the population ecology of guillemots, before I'd even seen one. Their objective was that I should try to understand and explain the massive decline in guillemot numbers that had also occurred in southern Britain over the previous two or three decades. I was excited at the prospect of solving this particular puzzle, but even more fired up by the sexual behaviour of guillemots.[28]

As an undergraduate just the year before, I had been introduced to the thrilling new field of behavioural ecology that, among other things, put a novel spin on promiscuity. Rather than dismissing it as some kind of biological aberration, the behavioural ecology approach suggested that it might be adaptive. I was fascinated by the possibility that Common Guillemots, despite being classed as 'monogamous', might sometimes engage in extra-pair copulations. In the winter before I started my field work on Skomer I came across Arne Nørrevang's study that, to my delight, confirmed that such promiscuity was prevalent among guillemots: 'One female,' he wrote, '. . . was mounted by three different males in twenty minutes.' I was ecstatic, and promiscuity formed the foundation for much of my subsequent research on guillemots and other birds.[29]

Dove-grey and doe-eyed, the fulmar petrel – the Northern Fulmar – is a beautiful mini-albatross. But looks can be deceptive, and this is a bird with a sharp bill, a rank smell and the unsavoury habit of projecting a stream of hot, oily vomit into the face of the unwary. It was the uncanny accuracy of the fulmar's projectile vomiting that gave rise to its name: foul mar (foul gull).

The fulmar petrel, which first bred on the Faroes in the mid-1800s (from Yarrell, 1843).

In Lucas Debes's day the fulmar was unknown in the Faroes. Even in the late 1700s George Landt, another Danish clergyman stationed there, wrote that the fulmar 'is known only to those who fish a great way out to sea'. The species first started to breed on the Faroes in the mid-1800s, as its population – originally just a single colony in Iceland in 1640 – began to radiate outwards across the North Atlantic in response, it is thought, to the food unintentionally supplied by commercial whaling and, later, commercial fishing. Fulmars are scavengers, eating anything dead on or in the sea. A fulmar I captured in Labrador regurgitated part of a dead cat, which I presume had been lost at sea.[30]

In the early 1800s, before starting to breed on the Faroes, the fulmar was considered a bird of ill omen there. This idea may have had its origins on St Kilda, a similarly isolated archipelago some 275 miles (440 km) to the south, where in the late 1600s Martin Martin described it as a 'sure messenger of evil tidings'. This was because the fulmar's appearance in November (i.e. outside the breeding season) was 'always accompanied with boisterous west winds, great snow, rain or hail'. As the fulmar settled to breed in the Faroe Islands in ever-increasing numbers – first doing so in 1839 – the locals had reasons to both curse and bless this extraordinary intruder.

Invader it was: ousting the guillemots – one of the favourite birds of the Faroese – from their breeding ledges, splattering them with its evil oily vomit that, like commercial mineral oil, destroys their waterproofing. It wasn't until the early 1900s that the Faroese people started to eat their unwelcome guest, a curious inertia, given that on St Kilda the residents were said in the late 1600s to 'prefer this, whether young or old, to all other [seabirds]; the old is of a delicate taste, being a mixture of fat and lean, the flesh white . . . the young is all fat'. Not everyone agreed. In the Disko region of Greenland, fulmars were considered 'too foul to eat; but at Umanaq they eat young birds'. Someone subsequently 'tried fulmars at both places and found a remarkable difference in their taste', a difference that was due – it was supposed – to the fact that the fulmars at Disko fed on whale excrement.[31]

How things have changed. There are now thought to be over half a million pairs of fulmars breeding on the Faroes and for the last 120 years they have been harvested and consumed in huge numbers with a contradictory mix of relish and disdain. To some Faroese, the fulmar harvest is still considered a 'service' to minimize their effect on the guillemots.

Like most other seabirds, fulmars lay a single large egg, averaging around 100g, so roughly twice as heavy as a typical chicken egg, but also relatively large for the size of the bird. Not only that, a fulmar egg contains a relatively large yolk, so each one is a nutritious prize. Acquiring them from the grassy slopes and rocky ledges of the gigantic cliffs demands the same technique that was used to collect eggs from the guillemot breeding ledges. Teams of ten or more men carry huge hemp ropes to the cliff tops, where they hammer in two wooden stakes to secure one end of the rope. Then, wearing a robust sheepskin harness attached to the rope, and handcrafted sheep's wool non-slip 'slippers', one man is lowered down to the ledges where the fulmars nest. Once among the birds the man steps out of the harness so that he has more freedom to move around – a foolhardy and dangerous game of machismo that more than once has resulted in a fatal fall. In 1857, no fewer than eight men died during the egg harvest.[32]

The fulmar eggs are gathered into buckets, which when full are hauled back up to the cliff top to unload. The eggs are then shared among the participants and other island residents. As James Fisher states in his monograph: 'There are no two opinions about the taste of the fulmar's egg: it is very good.'[33]

Compared with the other seabirds, the fulmar chick is a slowcoach, remaining some six weeks in the nest, during which time it is fed a concentrated oily mix of marine offal and plankton by its parents before fledging. The result: one fat chick, weighing more than each of its parents, with almost half its bodyweight fat – a food reserve – a tactic that has evolved to see the young bird through its first few weeks of independent life. But, of course, it is that fat that makes the fulmar chick such a tasty dish.[34]

The current harvest of recently fledged fulmars in the Faroes is huge and harsh. At fledging the young fulmar is so fat that it does little more than flop down onto the sea from its nest site on the cliffs, where it bobs around like a blob of lard floating in a washing-up bowl. Using high-powered boats and a long-handled net, the locals scoop them up in their hundreds, or thousands. I was told of one boat that collected 10,000 chicks in a single season. There are no legal limits and some 40–80,000 fulmar chicks are sold each year in the islands' capital, plus those retained for home consumption. The birds are ruthlessly processed as they are brought aboard: heads ripped off and the oesophagus and gizzard dragged out (to avoid their oily contents) and discarded. Plucked and singed to remove the down with a blow torch (which adds a certain smokiness to the taste), gutted and eventually roasted in someone's oven, the fatty young fulmar is served and eaten with potatoes. The general consensus among the Faroese I spoke to is that it is delicious. Outsiders, however, are less convinced.

Harvesting, preparing and eating fulmars has come at a

cost – retribution, perhaps – for in the 1930s no fewer than 179 Faroese people died from psittacosis as a result of handling, plucking and preparing young fulmars. Pregnant women were especially vulnerable, and most of those who became infected died. The original source of the psittacosis bacterium (*Chlamydia psittaci*) was though to have been shipments of parrots from Argentina that in the winter of 1929–30 caused a widespread epidemic of psittacosis across Europe and the United States. The Faroese authorities banned the hunting of fulmars in 1939, but the hunt was reinstated in 1954 after additional precautionary measures were introduced. The incidence of psittacosis subsequently declined to one or fewer cases per year, none of which has proved fatal.[35]

Just as we might expect, a place as remote as the Faroes, and with such long dark winters, should host a rich and varied folklore. On the island of Mykines, on the west side of the archipelago, the women welcome back the gannets each January by congregating on the cliff tops and waving white handkerchiefs – symbolic of the birds themselves.

The final chapter of Lucas Debes's book, entitled 'Of Spectres and Illusions of Satan', helped to ensure a greater success than the book might otherwise have had. I have to say, I failed to understand much of that chapter, but it is essentially about the way the Faroese people dealt with the inexplicable, such as illness, dreams and visions. As a clergyman, Debes was convinced that the Faroes have always been 'an habitation of devils, a domicill for unclean spirits and a den of goblings [*sic*]' and sees the inexplicable as part of the battle between good

Welcoming the return of the Gannets to Mykines on 25 January 1925 (photo: Ullrich Bodo). Gannet chicks form an important part of the Faroes seabird harvest.

and evil; between God and the devil. He also recognized that the unfortunate state of the Faroese people was because these are 'flegmatic folks that do not much stir themselves' and because the atmosphere beside the sea is cold and moist. Apart from anything else, this gives one a sense of what Debes was up against in trying to 'save' his flock.[36]

Given the close link between birds and people in the Faroes, birds seem to have featured rather less in their folklore than one might have imagined. Debes recounts the ritual burning of ravens' beaks each year, but there's no real mystery since this was merely the culmination of a cull used to curb the ravens' attack on young lambs and weak sheep.

A tiny woodcut from Olaus Magnus's *History of the Northern Peoples* from 1555, showing a man delivering the heads of ravens he has culled to avoid being taxed for failing to deliver at least one head.

Debes, however, is intrigued by these birds:

> Amongst those ravens there are found some white, though few: but those that are half white and half black are fit, when they are taken young, and have the tongue string cut, to be taught to speak. I have made a notable experiment upon a young white raven, whose tongue string was cut; and yet I had no thought of teaching him: but calling usually in the morning upon my boy, whose name was Erasmus, and the raven continually in the morning hearkening to that word Erasmus, begun at last to call out *Erasmus*, before the chamber where the boy lay, forming its voice exactly after mine: the boy hearing it, answered, anon master; and therewith arose and came into the chamber, to know what I would have: but I telling him that I had not called him, he went to bed again but was again called in the same manner and was so deceived by the raven.[37]

Once he figured out what was going on, Debes says, 'I began purposely to teach it,' describing how the bird would repeat what he said, 'putting the syllables together, till it could at last speak out the whole word, as children do, when they learn to spell in the schools'. As Aristotle noticed (Chapter 3), and as recent research into the ability of birds to acquire both their natural songs and, in the case of corvids and parrots, human speech has confirmed, the process in both birds and children is very similar.[38]

Debes wasn't the first to comment on the white-speckled raven of the Faroes – the *hvitravnur*. It is mentioned in a ballad of birds dating from the 1500s, and Debes's tutor, Ole Worm, had two specimens, collected around 1650, in his cabinet of curiosities. The white-speckled raven is quaintly but beautifully illustrated by Didrik Sørensen. Once thought to be a distinct species, the pied raven was a Faroese speciality, a variant of the Northern Raven in which some of the plumage lacks pigment. Sadly, but inevitably, as museum collections started to expand in the 1800s, everyone wanted one (or more), and the last pied raven was shot in 1902. There are just twenty-six specimens in museums today.[39]

For many years Britain had its own equivalent of the Faroes, with inhabitants dependent on seabirds. The St Kilda archipelago 40 miles (64 km) west of the Outer Hebrides is in some ways more familiar than the Faroes, and with good reason. For over 200 years St Kilda was considered 'among the greatest curiosities of the British Empire' – where else in the British Isles were people so isolated and men such able 'cragsmen', scaling the vertiginous cliffs in search of seabirds? Where else in the British Isles did people live in such filthy, rural squalor, yet seem so content? The St Kildans were the ultimate noble savages that Victorian tourists came to gawp at as though they were freak-show exhibits. Like the Faroese, the St Kildans were reliant on seabirds, eating them fresh and drying them in their beehive-like stone cleits for the winter.[40]

It was Martin Martin who first made the St Kildans known to the modern world through the account of his visit in 1697: *A Late Voyage to St Kilda* ('late' here meaning recent). Setting off from the island of Harris in an open boat, it proved to be a long voyage, for as well as the sixteen stormy hours it took to reach St Kilda, on their arrival the seas were so rough that Martin and his colleagues were forced to shelter under the cliffs and a constant shower of gannet guano that 'sullied [their] boat and clothes' – for two days and a night before being able to set foot on Hirta, the main island.

If Martin wasn't inspired by Lucas Debes's description of the Faroes, published twenty-two years earlier, the Royal Society was, and in their quest for new knowledge it was they who encouraged Martin to 'survey the isles of Scotland'. They could hardly have found anyone better, for Màrtainn MacGilleMhàrtainn, as he was known locally, was a well-travelled, well-educated and well-heeled factor on an estate on Skye. A Gaelic speaker, Martin was able to converse with the St Kildans, who knew no other language.

When he and his shit-splattered colleagues finally staggered ashore, the St Kildans welcomed them and, as was their Christian custom, provided them – sixty men in total – with lodgings and food during their three-week stay. Even so, Martin and his colleagues were somewhat taken aback by their daily ration of a barley cake and eighteen Common Guillemot eggs. But they were hungry and, 'Our men upon their arrival eating greedily of them, became costive [constipated] and feverish.' As Martin said: 'The eggs are found to be of an astringent and windy quality to strangers, but it seems not so to the inhabitants, who are used to eat them from the nest.'

For Martin and his colleagues the consequences of this overegged diet was, he noted, diabolical. Some of the men had 'the hemorrhoid veins swelled', and Martin tells us that 'Mr Campbel* and I were at no small trouble before we could reduce them to their ordinary temper'. They treated themselves with a 'glister' (enema) made from 'the roots of sedges, fresh butter' and salt, which, being administered, had its wished-for effect. The inhabitants, Martin says, 'reckoned this an extraordinary performance, being, it seems, the first of this kind they ever had occasion to hear of'.[41]

Martin estimated that the number of eggs 'bestowed upon those of our boat' during their visit was 16,000, and 'without all doubt the inhabitants, who were triple our number [i.e. 180], consumed many more eggs and fowls than we could. From this it is easy to imagine, that a vast number of fowls must resort here all summer.' And vast indeed must have been the seabird populations at that time, despite centuries of human depredation.[42]

As with the Faroes, we have no idea of the numbers of seabirds on St Kilda in the late 1600s, but the populations of Atlantic Puffins, Common Guillemots, Northern Fulmars and Northern Gannets were clearly huge, with smaller numbers of Razorbills, Manx Shearwaters, Black Guillemots,

* The Reverend John Campbell from Harris, sent to administer to the St Kildans' needs, who accompanied Martin on the steward's annual visit.

Fowlers on St Kilda: photograph by George Washington Wilson from the 1880s (courtesy of University of Aberdeen).

Black-legged Kittiwakes, European Storm Petrels and gulls. The fact that the inhabitants of St Kilda could provide 16,000 guillemot eggs for Martin and his colleagues gives us some idea of their numbers.

Another indication of the abundance of seabirds was the harvest of feathers.

Like the Faroese, the St Kildans were held in feudal bondage by their landlords and their bailiffs:

> When the steward and his followers come among them [the St Kildans] to demand his rents etc., viz. down [feathers], wool, butter, cheese, cows, sheep, fowls [seabirds], oil [from fulmars] etc., they look upon that visit as no great advantage to them and they very much grudge what he carries away with him and that they must be all the year toiling for others.

The late-1700s fashion for feather beds meant that between 1793 and 1840 the annual quantity of feathers stripped from St Kilda's seabirds rose from 20 stone to 240 stone – the latter equivalent to 100,000 puffins.[43]

There is in Martin's account a brief mention of the land birds of St Kilda: 'Hawks extraordinary good, eagles, plovers, crows, wrens, stone chaker [stonechat?], craker [Corncrake?], [and] Common Cuckoo.' He continues: 'this last – the cuckoo – being very rarely seen here, and that upon extraordinary occasions, such as the death of the proprietor . . . or the arrival of some notable stranger'. When Martin laughed at this the St Kildans 'wondered at his incredulity', insistent that it was true and had in fact happened just before his own recent arrival.[44]

The St Kildans' cuckoo folklore contrasts strikingly with their much more reliable knowledge of the seabirds. The St Kildans that Martin met and interviewed knew a great deal about the birds both on which their lives depended and for which they were risked as they descended the cliffs. They knew where and when they bred; they knew how many eggs they laid in a season; whether they would produce replacement eggs if their first was taken; they knew about the effect the wind had on the birds' presence at the colonies, and the way they could foretell the future by anticipating bad weather.

The St Kildans abandoned their island home in 1930 as a result of dwindling population, a succession of crop failures and a general sense of despair. Like the premature death of a rock star, the St Kildans' departure for the Scottish mainland fixed them permanently in the public imagination and has been responsible for the continuous memorializing of them. Significantly, by leaving when they did, the St Kildans missed the slow crash in seabird numbers experienced by the Faroese and, with it, the condemnation of their way of life.

Old images of the Faroese, St Kildans or people from other North Atlantic communities catching and killing seabirds or with enormous piles of eggs can seem quaint – windows onto a lost way of life in which people depended on birds for food. It is all too easy to romanticize the past. The lives of these people were harsh and they risked life and limb in their quest for sustenance. But with our current values shaped by rapidly declining bird populations, those images now seem unacceptable. In a similar way the Faroes' grind – the traditional pilot-whale hunt – in which several hundred long-finned pilot whales are killed each year, was once seen as a charming, if violent and bloody, local tradition. But with near-global opposition to killing whales since then, the grind is now the target of fierce protests. The protesters argue that it is no longer necessary because Faroese society now enjoys a much higher standard of living than previously, with alternative sources of protein. And possibly the main point is that the killing of such intelligent, social animals is cruel. The killing of seabirds – puffins, for example – in the Faroes and Iceland doesn't generate as much debate, but for some it is emotive, not so much because they are social and intelligent, but because they are cute clowns of the sea. That's irrational, given that there's even less opposition to the killing of the less-cute guillemots or fulmars.

The Faroese counter the grind protesters by referring to the cultural importance of hunting; the fact that the meat is natural, has led a free life and is sustainable: 'We don't want to depend on imported food and eat animals kept in captivity all of their lives!'[45]

The hunters I spoke to in the Faroes also stressed the need to retain their hunting skills – for both whales and birds. The recent rapid rise in the standard of living on the Faroes is a result of salmon farming, from which many different sectors of society benefit. If, they argue, the bottom were to fall out of fish farming, they would be forced to fall back on traditional sources of protein, and this would be difficult if they allowed their hunting skills to die. This is exactly the response we might expect from a society that has paid a high price, historically, for its isolation and periodic famines.

With concern about the global ongoing loss of biodiversity and bird populations, it is easy to stir up an emotional response to any 'unnecessary' killing. It is easy to sell newspapers by showing the bloody realities of the grind or the Faroese fulmar hunt. It is much less easy to take an objective stand and present both sides of the argument.

On the face of it, this seems like a contest between the Faroese people and the protesters, but it is much more complex than that, for there is a third player in this messy dispute: the rest of the industrialized world. The pilot whales that the Faroese continue to catch are now so contaminated with mercury, DDT and other industrial chemicals that they are no longer fit for human consumption. Some die-hard Faroese continue to eat whale meat, but pregnant women are advised not to, unless they are prepared to risk their children suffering developmental and mental-health issues later in life.

Over the same period that toxic chemical levels have been increasing in whales, guillemots and puffins have suffered at the hands of climate change. There are no longer sufficient guillemots to exploit at the colonies, and the failure of the puffins to breed successfully has resulted in a self-imposed moratorium on hunting. If, the Faroese say, chemical pollution and climate change are so evident in their remote and tiny part of the world, what can it be like elsewhere? Neither toxic chemicals in the marine environment nor climate change are the fault of the Faroese. The Faroes now represent a microcosm of the rest of the world.[46]

8. The End of God in Birds: Darwin and Ornithology

> Extinguished theologians lie about the cradle of every science as the strangled snakes beside that of Hercules; and history records that whenever science and orthodoxy have been fairly opposed, the latter has been forced to retire from the lists, bleeding and crushed, if not annihilated; scotched, if not slain.
>
> Thomas H. Huxley (1870)

As a research student in the 1970s, I travelled, with my friend and fellow biologist Nick Davies, by train, bus and on foot from Oxford to Downe in Kent, to see Charles Darwin's home. My overriding memory of that day is of the inebriated caretaker showing us round the musty, uncared-for house and his attempts to persuade us, towards the end of our visit, to join him for a drink in the pantry. I also remember the enormous glass case of stuffed hummingbirds in the entrance hall as we walked into the house. With spread wings and tails, these tiny, static jewels appeared to be fixed in time. A brazen piece of archetypal Victoriana, the hummingbird case was not part of the original Darwin furniture, but added later, coincidentally providing a link between Darwin and the ornithologist John Gould.

By recognizing that the finch specimens Darwin brought back from the Galápagos were all related, Gould nudged Darwin's thoughts ineluctably towards natural selection – rather than God – as the mechanism by which transmutation caused species to change over time. In 1838, not long after Darwin's return, and with no mention of transmutation, Gould and he were on good terms. Darwin even sent him a 'Dram-bottle' (a silver-cased compass) for his forthcoming trip to Australia, and Gould named the 'new ostrich' (actually a rhea) Darwin had discovered in South America *Rhea darwinia.*[1] In later life, Gould – by then the ultimate ornithological celebrity as a result of his magnificently illustrated volumes on birds – had manoeuvred himself into the position of world hummingbird expert, which meant hummingbird killer. Ever since my first visit to Down House, that display of dead hummers in Darwin's hallway has haunted me. It was to be another few years before I was thrilled by the sight of a live hummingbird, a Ruby-topaz, in Tobago.[2]

Twenty years later, after Darwin made public in the *Origin* his idea that natural selection and not God had shaped the natural world, he and Gould fell out. It is hard for us today to understand the nerve-shattering nature of Darwin's announcement. *On the Origin of Species* ripped apart the comfortable, reassuring bubble that John Ray had created 150 years previously that so elegantly blended God and the natural world. Ray's natural theology posited that God was responsible for the near-perfect match between an animal and its environment and encouraged readers to seek evidence for God through the study of nature. In doing so, *The Wisdom of God* elicited the beginnings of field ornithology and the study of ecology and animal behaviour.

Among those inspired by Ray, the best known is Gilbert White. His forty-year tenure in Hampshire as Selborne's parson during the 1700s provided him with unparalleled knowledge of the local fauna and flora, and his *Natural History and Antiquities of Selborne* in turn inspired others to write similar books. Thought of as the 'first ecologist' and (another) 'father of ornithology', White was both perceptive and pragmatic, although I have to admit I was shocked to read that he had killed twenty-four bullfinches that 'lay very hard on the Cherry-trees & Plum-trees, & had done a great deal of mischief'.[3] However, it was through having shot birds that White was the first to distinguish the three leaf warblers – the Common Chiff-chaff, Willow Warbler and Wood Warbler – from one another. Never out of print, *Selborne* is a beautifully written, ingenious blend of scientific and emotional responses to nature.[4]

In January 1789 White noted that a neighbour, while crossing Wolmer forest, had 'found a large, uncommon bird fluttering in the heath, but not wounded, which he brought home alive'. The unfortunate bird was a Great Northern Diver, a long way from its normal aquatic habitat. White examined it, noting how 'Every part and proportion of this bird is so incomparably adapted to its mode of life, that in no instance do we see the wisdom of God in the Creation to more advantage.'[5] Gilbert White was a great admirer of Ray and the *Ornithology of Francis Willughby* was his standard reference book on birds.[6]

White's empathy for the natural world helped to make him an iconic figure. But *Selborne* was not an immediate success. Initial reviews were positive, but it was not until 1827 and the publication of ornithologist William Jardine's edition, thirty-four years after White's death, that *Selborne*'s star began to rise. And it had little to do with that particular edition and much more to do with a surge in publishing and reading among the British public. By the 1820s books were cheaper to produce, and a much greater proportion of the population was now literate and had sufficient cash to buy books and enough leisure time to read them.[7]

What the public was so voraciously reading was science. White's comment on the Great Northern Diver anticipates the burgeoning interest in natural history and natural theology. Through the 1800s British society became increasingly religious, and what better way to serve God than by celebrating his creations? *Selborne* provided the perfect guide to the close observation of nature, and in particular the scrutiny of what had previously been dismissed as inconsequential or commonplace. There had long been a market for tales of exotic wildlife in distant lands, but White demonstrated that by looking at the birds and other wildlife on one's doorstep, everyone could engage with nature. The act of observing and recording became an act of devotion: a way to serve God. While this seemed like a simple continuation of Ray's natural theology, there was a difference. Ray's passion for the natural world was a quest for evidence for the existence of God. By the early 1800s, God's presence was taken for granted, and the observation of nature had become a 'reverential practice'.[8]

One of those enthused by White was the Reverend Leonard Jenyns, curate at Swaffham Bulbeck, in Cambridgeshire, who had read *Selborne* 'with avidity' at school, 'It being so entirely in keeping with my own fondness for natural history and out-of-doors observations on the habits of birds and other animals'. At the age of thirty, Jenyns was Captain Robert Fitzroy's first choice to serve as his companion on the *Beagle* voyage, but luckily for us all, Jenyns declined, partly due to his fragile health but also because he did not want to abandon his parishioners. Instead, Jenyns, together with his brother-in-law the Reverend John Henslow, recommended Charles Darwin.[9]

Jenyns and Darwin shared a fascination for beetles, and they first met at one of Henslow's Friday evening soirées. Initial impressions, however, weren't good, and Darwin thought

Jenyns very old school. A retiring, unsmiling bachelor at the time, whose only real interests were natural history and collecting insects, Jenyns – like his hero Gilbert White – kept a naturalists' calendar, a diary of seasonal natural-history events. Eton- and Cambridge-educated, Jenyns was a well-respected naturalist, whom others had tried – unsuccessfully – to persuade to stand for a zoological professorship. Darwin invited Jenyns to Christ's College to see his collection of beetles, and generously gave the older man a few specimens. Darwin later wrote that he had initially disliked Jenyns for 'his somewhat grim and sarcastic expression', but added, 'and it is not often that a first impression is lost; but I was completely mistaken and found him very kind-hearted, pleasant and with a good stock of humour'.[10] They remained good friends throughout their lives.

Using *Selborne* as his guide, Jenyns produced *Observations in Natural History* in 1846, the main objective of which, as its title suggests, was to encourage his readers to *look*, to observe. As he said: 'although not the true end of the science of natural history, it is nevertheless a means to that end'. And that is what his book is about, *observing*, as White himself had done; observing, for example, the dates when the first song of a particular species was heard in the spring, or the dates when the migrant birds first arrived and departed.

Jenyns's book also contains a collection of natural-history anecdotes:

> 15 February 1829: Siskins everywhere.
>
> Spring 1841: a completely white rook was shot near Cambridge . . . [now] preserved in the Museum of the Cambridge Philosophical Society.

> 18 June 1827: To-day, whilst mowing the hay in the field in front of the vicarage, we found the nest of a landrail [Corncrake] containing seven eggs.

Jenyns included the tale of a solitary female 'coddymoddy' (Common or Mew Gull) that lived in the Provost of King's College's garden for many years, and regularly laid (infertile) eggs. In 1844 the gull's eggs were replaced with the fertile eggs of a duck, and a duckling subsequently reared by the gull. From the 1820s such natural-history 'trivia' started to become extremely popular and other writers were quick to capitalize on this interest.[11]

At the age of eight, already keen on birds, I was taken with two of my cousins to the home of my great aunt and uncle, Ella and Harry Llewellyn Bassingthwaite, in Norwich. Theirs was a dark, austere house – as I imagine Leonard Jenyns might have lived in. It was full of books, for Harry was a schoolteacher, while Ella, poor woman, was no more than his housekeeper locked inside a loveless marriage.

Towards the end of our visit we were taken into Uncle Harry's study and ushered towards a floor-to-ceiling bookcase full of natural-history volumes. To my utter amazement, we were invited to select a volume to take home as a gift. The one I chose was the Reverend John Wood's *Natural History* of Birds, published around 1851: impressively weighty and full of images. I was thrilled, and that book became a kind of bible for me. I read the stories about pet birds and admired the exotic range of species. To my present shame, I annotated the text (in pen!), recording when I'd seen each species, using coloured inks to enliven the black-and-white engravings. The binding eventually disintegrated through overuse, and decades later I bought a pristine copy, partly to assuage my guilt for defacing Uncle Harry's gift. The two now sit side by side in my own floor-to-ceiling bookcase.

Wood's *Natural History of Birds* embodies the increasing enthusiasm for learning about birds in Victorian times. Rich in images of species from across the world, with information culled from a vast number of sources, Wood's volume is designed to enchant. There's little that's original, but you cannot fault Wood's industry: descriptions of 550 species, 780 pages of text and a third of a million words. The sheer labour of scouring the literature and putting all this together in a coherent manner was monumental. It is an odd mix, though, intended to both inform and entertain, with plenty of emphasis on the unusual, including a big section on hummingbirds. Wood's taxonomic arrangement of species seems bizarre by our present standards: who today, for example, would place the Meadow Pipit alongside the White-faced Epthianura (an Australian honeyeater)? Many of the species illustrated and described would have been ones his readers can never have heard of and Wood himself could never have seen, but his style was captivating, as in this piece on toucans:

> Grotesque as is their appearance, they have a great hatred of birds who they think to be uglier than themselves, and will surround and 'mob' an unfortunate owl that by chance has got into the daylight with as much zest as is displayed by our crows and magpies at home under similar circumstances.[12]

It is odd that Wood considered toucans 'grotesque', and stranger still that he suggests that their mobbing behaviour might be motivated by a kind of aesthetic elitism. Today, mobbing is considered a way of driving away a potential predator. But by giving his birds human sentiments Wood captured his readers' hearts.

Wood gave up his curacy, presumably on the strength of his book's success, to write and lecture full time. An accomplished speaker, he delighted his audiences with what seemed like effortless, impromptu sketches, exquisitely executed in coloured pastels on a dark canvas. They may have seemed effortless, but they were meticulously rehearsed, for Wood was a showman. Over the years he published more than a dozen titles on a range of natural-history topics. And they were extremely successful: his *Common Objects of the Country*, published in 1858, sold 64,000 copies. By comparison, Darwin's *Origin* sold just 10,000 copies over a similar period. Despite his spectacular sales, Wood made little profit and was perpetually hard-up. After his death, his unfortunate widow was forced to apply for support from the Royal Literary Fund.[13]

Who read the bird books by authors like Wood? We do not know for sure, but we can imagine it was those who were slightly better off, motivated to improve their knowledge, and in some cases their morals too. One thing we can be fairly certain of is that most of the readers of these books were armchair birders. They had little choice, unless, like the professional collectors, they went out with a gun to shoot specimens. Birds were unusual in this respect. By comparison, the Victorian fashion for seashore animals or ferns allowed enthusiasts to get out onto the shoreline or the field and acquire their own specimens, stripping bare the littoral zone or humid ferny glades. Being an armchair ornithologist allowed readers to enjoy the biology, the stories and the images on the books' pages, but for some it was all a bit remote and they craved a closer relationship with their subjects.

Birds in captivity provided just such an intimacy as well as an opportunity for the sort of close observation that is almost impossible with wild birds. What's more, caged birds are utterly dependent on those who own them, and caring for captive birds created a sense of moral responsibility in their owners. In particular, having birds nesting and rearing young in cages – notably canaries, for they were the ones most likely to do so – provided a microcosm of the human condition: a devoted couple, working in harmony to rear their family. What better life lesson could there be? Even the death of a caged bird (and there would have been many) could be instructive, especially for children.

The keeping of songbirds in captivity was pervasive in Victorian England, where it was fostered and encouraged by, among others, the Society for the Promotion of Christian Knowledge:[14] 'God created . . . every winged bird according to its kind. And God saw that it was good. God blessed them and said, "Be fruitful and . . . let the fowl multiply on the earth." '[15]

The fashion for keeping and caring for birds in captivity – so much more civilized than cock-fighting or bull-baiting that had been popular in previous times – was seen as a step in the right moral direction. The author of *A Treatise on British Song-birds*, Patrick Syme, called the keeping of cage birds an 'innocent recreation', saying: 'Does not the parental conduct of these little creatures display an instinct that seems nearly allied to [our] moral feeling? And may not man follow their example with honour to himself and benefit to his species?'[16]

It was commonly assumed that birds in captivity sang for 'joy', or to please their owners: 'Singing birds are so pleasant a part of Creation . . . they were, undoubtedly, designed by the great Author of Nature, on purpose to entertain and delight mankind.' Just as they had 'gladdened the hearts of Noah and his family, when they sailed over the wild wastes of waters in his ark'.[17]

More knowledgeable, or less deluded, bird-keepers like Johann Bechstein, the German author of the best-selling bird-keeping book throughout the 1800s, were well aware that birds were not singing for joy, nor for their owners' pleasure. They realized that male birds sang to attract a partner and to defend a territory.[18]

As Britain's Industrial Revolution rumbled grubbily onwards, bringing wild birds into people's homes allowed poor, humble urban families to delight in their company. The trade in wild birds and the sheer variety of species claimed suitable for life in confinement seem extraordinary by today's, or indeed any, standards. Among the ninety-two species listed in one book, the Common Kingfisher, White-throated Dipper and Eurasian Wryneck have requirements for life that are so specific, I cannot imagine that they survived very long in captivity.[19] But just like the tadpoles collected by children today, wild birds were expendable and easily replaced – for there was, it was assumed, an infinite supply.

Not everyone approved of keeping birds that were taken from the wild (with the exception of canaries), but enthusiasts had a biblical response: 'A common objection, which some austere men (pretending to more humanity than the rest of their neighbours) make against the confining of songbirds in cages . . . as thereby abridging [depriving] them of their natural liberty . . .'[20]

Canaries were extremely popular: attractive, vocal and easy to rear. First imported from the eponymous islands in the 1500s, over the next few hundred years enthusiasts ingeniously transformed this nondescript green songster into the now familiar yellow variety. These birds were particularly popular with the working class and, as a consequence of their efforts in artificial selection, they not only altered the bird's colour but also changed its shape and size, with distinct varieties named after the regions from which they emerged, including Norwich, Yorkshire, Fife and Gloucester.

Writing about the benefits of keeping canaries in the 1830s, the Reverend Francis Smith said that children: 'may silently learn those invaluable lessons of kindness, and love and patience, which shall fit them for the trials of after life, and, it may be, be imprinted on their hearts for ever'. He adds: 'Who can say what stimulus and encouragement such an aviary might not often afford to the study of every department of natural history, which but for it might never be undertaken?'

The Victorians held contradictory views of birds (and other animals). On the one hand, birds were thought to be distinct from humans, with none of the same feelings as ourselves, which meant that they could be exploited and mistreated with impunity. On the other hand, many owners liked to think that their pet birds shared the same senses and emotions as them: 'They are grateful and affectionate little creatures. Some have even been known to pine and die, when deprived of those they were attached to.'[21]

We might scoff at that today, but most dog owners would

readily admit to such a bond, and the feeling between a person and their pet bird can be equally strong. The hard-nosed science of the close relationship we might have with a bird or other animal is called 'imprinting': a pre-programmed disposition of an animal, usually at an early age, to fixate itself to another individual – usually its parent. This is the mechanism by which young animals identify the species with which they should later mate. It doesn't mean they will want to pair up with a parent, simply another individual of the same species as their parent.

There are numerous examples of such cross-species imprinting, and one of the earliest studies showed that if you place the eggs of a Bengalese Finch (aka White-rumped Munia) into the nest of a Zebra Finch and allow them to rear the young, the Bengalese chicks will attempt to pair up with a Zebra Finch when they become sexually mature rather than with one of their own species. This is because, while in the nest, the young Bengalese Finches have imprinted on their Zebra Finch foster parents. Imprinting on whoever rears you is a good rule because in normal circumstances young birds are reared by their parents. Our understanding of the science of the bonds between people and birds takes nothing away from the affection that human individuals and pet birds feel for each other: it can be very real.[22]

I can identify with all of this. When I was growing up in Leeds in the 1950s, bird-keeping was widespread. As well as watching birds in the wild, I was desperate to get close to them. My father built me an aviary in the garden, and it was my childhood experience of keeping and watching these birds that was later to play such an important role in my scientific career. Among the many birds I had were Zebra Finches, and it was my experience of breeding them that later allowed me to explore the consequences of avian infidelity.[23]

The publication of Darwin's *Origin* on 24 November 1859 resulted in a very public dispute the following June, in which the Bishop of Oxford, Samuel Wilberforce, batting for the Church, was trounced by Darwin's ferociously loyal friend Thomas Henry Huxley. One could be forgiven for assuming that, following that debate, God ceased to be part of our relationship with birds. But it wasn't that simple.

In preparation for the confrontation in Oxford, Bishop Wilberforce was coached by Richard Owen, Britain's top zoologist and palaeontologist, but also an unctuous Establishment figure who hated Darwin with a vengeance. Owen had read the *Origin* and had written several anonymous mean-spirited reviews of it, so he was well placed to brief the bishop. On the day, however, Wilberforce, who was known as an outstanding orator, behaved like an overconfident tutee who, without having prepared sufficiently, tries to blag his way through a tutorial with clever but vacuous turns of phrase. The bishop, also known as 'Soapy Sam', had not read the *Origin* and Huxley made mincemeat of him.

The idea of organisms changing or transmutating over time was not new with Darwin. He had several predecessors, including Jean-Baptiste Lamarck in the early 1800s and Robert Chambers, author of *Vestiges of Creation* in 1844. Darwin's key idea, however, first conceived in the 1830s and boosted by John Gould's insights relating to the Galápagos finches, was that transmutation occurred through the process of natural selection, mediated by limited resources. Darwin's idea was inspired by reading Thomas Malthus's work on population and the way a 'struggle for existence' resulted in favourable variations being preserved and unfavourable ones destroyed.[24] Darwin looked at how poorly Lamarck's and Chambers's ideas were received by the public, and held back on publishing. Denying any role for God, Darwin knew, would be a bitter pill for people to swallow. The time was not right and he waited twenty years, during which time he accumulated more evidence and refined his arguments. With hindsight it is obvious, as Darwin knew, that natural selection was a far better explanation for the traits of animals and plants than God and natural theology, but denying a role for God would be like touching a raw nerve.

It was receiving Alfred Russel Wallace's letter on 18 June 1858 that finally spurred Darwin to put pen to paper and produce what he called an 'abstract' of his idea and which would become the *Origin*. Wallace had spelled out a mechanism for evolutionary change, essentially the same as Darwin's natural selection.

Like lots of good ideas, Darwin's and Wallace's concept of natural selection was remarkably simple. So simple in fact that T. H. Huxley exclaimed: 'How extremely stupid not to have thought of that!' But, simple as it might be at one level, it proved difficult for many people to grasp. I remember at school struggling to understand it, largely because my teachers presented it in a muddled way. It wasn't until I was an undergraduate when natural selection was conceptualized as operating on individuals, rather than populations or species (Chapter 11), that it all became clear.[25]

To deny God in Darwin's day was to deny civilization. As a result, evolution by natural selection threatened to wrest power from the Church of England. It says a lot for Darwin's belief in his own idea, and the evidence he had found for it, that he was, eventually, happy to present it to the world. And, by doing so, prepared to risk alienating his deeply religious wife Emma.

Darwin's small band of loyal followers included T. H. Huxley and Joseph Hooker, with whom he had shared his ideas prior to publication, and it was largely thanks to them that the *Origin* survived the onslaught of opposition. Darwin also received support from the ornithologists Canon Henry Baker Tristram and his friend Alfred Newton, and some half-hearted support from the Reverend Jenyns, who, despite their long friendship, could not 'go all the way' with Darwin.[26]

Among Darwin's non-scientific supporters, one of the most influential was the writer the Reverend Charles Kingsley, later author of *The Water-Babies*. Knowing he was sympathetic towards science, Darwin sent Kingsley a pre-publication copy of the *Origin*. In response, Kingsley told him that the *Origin* awed him: 'if you be right I must give up much that I have believed'. For Kingsley, natural selection was no 'obstacle to faith: God didn't just make the world, he made the world make itself'.[27]

I have a soft spot for Charles Kingsley, and not just because he was unusual among the clergy in supporting Darwin's revolutionary ideas, but also because of an incident in my childhood. On a day out from our home near Leeds, my parents took my brother and me to Malham Cove in Yorkshire, a popular beauty spot made famous by artists such as James Ward and J. M. W. Turner. Climbing down the steep hillside towards the beck at the bottom, I looked down towards a tiny cottage, where I could see a woman in the garden, and I commented to my parents that this reminded me of a scene in Charles Kingsley's *The Water-Babies*. Surprised by the coincidence, my parents told me that this was exactly where Kingsley had been inspired to write his story. I have no idea how my parents knew that, and I can only presume that *The Water-Babies* had been read to me at school.

Kingsley was enthusiastic about the natural world, and sometimes imagined himself as a bird, on one occasion even making a drawing of himself as one. At school he had been tutored and inspired by the Reverend Charles Alexander Johns, who in 1862 produced the extremely popular *British Birds in Their Haunts*. At Cambridge, Kingsley was taught by Adam Sedgwick, who had earlier instructed the young Charles Darwin. As a schoolboy, Kingsley had been nesting with his friends and retained an interest in birds throughout his life. Ordained in 1842, he subsequently became chaplain to Queen Victoria and private tutor to her son the Prince of Wales, who would succeed her as Edward VII. Kingsley later served as canon of Chester Cathedral, eventually rising to the same position at Westminster Abbey. Despite his success and social standing, Kingsley was emotionally fragile. A polymath, radical and social reformer, he helped to launch natural-history societies and enthusiastically supported the educational role of museums. During his courtship of his wife Fanny, Kingsley sent her illustrations of his sexual fantasies, including one in which he imagined them together in a perpetual coital embrace, himself as a bird, bearing them heavenwards on his wings.[28]

Kingsley was keen on hummingbirds, and they were uppermost in his mind when in December 1869 he and Fanny set sail from chilly England to Trinidad in the warm West Indies to fulfil what he called 'a forty-year dream'. As guests of Sir Arthur Gordon, the island's governor, they

Charles Kingsley's image of himself as a bird carrying his beloved partner, Fanny, in a coital embrace (from Chitty, 1974).

revelled in the wonders of the tropical vegetation for seven weeks, but were disappointed by the lack of birds:

> Would that there were more birds to be seen and heard! But of late years the free Negro . . . has held it to be one of the indefeasible rights of a free man to carry a rusty gun, and to shoot every winged thing. He has been tempted, too, by orders from London shops for gaudy birds – humming-birds

> especially. And when a single house . . . advertises for 20,000 bird-skins at a time, no wonder if birds grow scarce.[29]

Charles Kingsley sits in startling contrast to another clergyman, the Reverend Francis Orpen Morris, an extremely successful Victorian popularizer of ornithology. The success of Morris's *History of British Birds* lay as much in the writing as in Benjamin Fawcett's woodblock colour prints of Francis Lydon's beautifully crisp paintings. The contrast with the Reverend J. G. Wood's dark, slightly muddy engravings was all too evident, and Lydon's images still look wonderfully fresh today. And the text, well . . . like several previous authors of popular bird books, Morris had scoured the literature for what others had written, amassing a large number of anecdotes and observations, including this fabulously unlikely one:

> A French surgeon at Smyrna, wishing to procure a [White] Stork, and finding great difficulty, on account of the extreme veneration in which they are held by the Turks, stole all the eggs out of a nest, and replaced them with those of a hen. In the process of time the young chickens came forth, much to the astonishment of the Storks. In a short time the male went off, and was not seen for two or three days, when he returned with an immense crowd of his companions, who all assembled in the place, and formed a circle, taking no notice of the numerous spectators, which so unusual an occurrence had collected. The female was brought forward into the midst of the circle, and after some consultation, the whole flock fell on her and tore her to pieces . . .

The implication here was that by producing bastard offspring the female had behaved dishonourably and had to be punished. Moralizing stories like this, however improbable, made Morris's books seductively attractive.[30]

Morris also used his books to rant against Thomas Malthus and Harriet Martineau (a friend of Darwin's brother Erasmus), presumably for their shared heartless and objective views that, as the human population grew and outstripped its food supplies, it would be the poor who would suffer most. Morris also lashes out at novelist Maria Edgeworth, whose stories, he felt, lacked 'a pervading religious principle'.[31]

An irascible and complex man violently opposed to Darwin's ideas and who regarded T. H. Huxley as a vile vivisectionist, Morris also was anti-feminist, but an early supporter of conservation. His son Marmaduke Morris, writing in 1897 about his late father, said:

> Of all the clergy of the Church of England who made public their opinions [on Darwin] there was probably not one who wrote at greater length [in pamphlets, periodicals and papers], more outspokenly, vehemently, and decidedly . . . Nothing in recent times had done more to unsettle the minds and the religious opinions and beliefs . . . than . . . the unwarrantable conclusions that had been drawn from the writings of Darwin.

As Marmaduke recalled, Morris Sr was unable to entertain the idea of a common ancestor, nor of natural selection. 'Where,' he asked, 'is the setting forth of the doctrine of evolution in the Book of Genesis? If Darwinian theory be true, then the Bible is untrue.' He went on:

> [Darwin's] mistake, the one great mistake, the one great cardinal error, as it appears to me, which runs through the whole of his work, is in supposing that because many mere

> varieties had their origin in one common ancestor . . . all distinct species are to be similarly accounted for. His whole argument is a *non sequitur.*[32]

Morris's view of Darwin reminds me of the zoologist Henry Gosse, as told in Gosse's son Edmund's 1907 biography *Father and Son.* A great popularizer of science, Henry made the seawater aquarium a Victorian sensation, and after an eighteen-month sojourn in Jamaica produced an excellent account of the birds there. As an ardent member of the evangelical Plymouth Brethren, Henry Gosse 'determined, on various grounds, to have nothing to do with Darwin's terrible theory, but to hold steadily to the law of the fixity of species'. Edmund, who to his father's dismay rejected God, described his upbringing as feeling 'like a small bird in a cage, imprisoned in a religious system'.[33]

For several centuries the Common Cuckoo had been a problem for those who believed in an all-knowing, benevolent God. What kind of benevolent divinity creates a creature that, as Aristotle first noticed, dupes another into rearing its offspring? What kind of God designs a tiny chick that kills its foster siblings? Surely, this requires some kind of explanation. John Ray, writing in the 1670s, says that he was stunned by the cuckoo, whose behaviour 'seems so strange, monstrous and absurd, that . . . I cannot sufficiently wonder there should be such an example in nature; nor could I believe such a thing had been done by nature's instinct, had I not seen it with my own eyes'. Yet later, in his *Wisdom of God* – his handbook of natural theology – he makes no mention of the cuckoo, avoiding the need to explain why God might have manufactured such a monster. Gilbert White simply called the cuckoo's behaviour 'an outrage on maternal affection', adding that, 'had it been described only of a bird of the *Brazils* or *Peru*, it would never have merited our belief'.[34] Even with his long and detailed account of this species, James Rennie dodged the issue entirely in his three little bird books published in the 1830s, books that were otherwise well researched and well written. Rennie's publisher, however, was the Society for the Diffusion of Useful Knowledge. Their aim was to make scientific topics accessible to the working classes, and Rennie seemed to have done this admirably. And while the society claimed to be secular, it vigorously supported a natural-theology view of the world, as did Rennie himself. Trained as a scientist, he made his living as a hack writer, with competitors referring to him as 'alphabet Rennie' for the various biological dictionaries he produced. His third volume concludes by saying that 'illustrations of the power, wisdom and goodness of the creator might be produced from the works of nature without end', and that the study of ornithology 'affords numerous examples of this'. He enumerates all the features of birds and explains each of them in turn as examples of God's wisdom – all except the cuckoo.[35] Similarly, in his *History of British Birds*, the Reverend Morris passes no judgement on the cuckoo chick's murderous nature.

The way the young cuckoo monopolizes the host's nest was first reported in detail by Edward Jenner in 1788 after watching a recently hatched chick eject the eggs of its Dunnock host, one by one. Jenner deals with the issue of how a Wise Creator could allow such cruelty, by suggesting that the death of the host's chicks was not without purpose because they provided a source of food for other animals. While not necessarily agreeing with Jenner's interpretation, most ornithologists accepted the accuracy of his observations. But Charles Waterton, an eccentric traveller and naturalist, considered Jenner's account 'preposterous', doubting that he had seen 'what he relates'. Despite his wide knowledge of birds, Waterton had clearly never seen a cuckoo hatchling, and could not accept that such a young chick had the strength to eject the host's eggs or chicks from a deep-cupped nest. He concludes by saying he would rather believe the story of young Hercules throttling two snakes in his cradle.[36]

Waterton's argument was echoed by John Gould, who in his *Birds of Great Britain* was keen to promote the idea of avian harmony, suggesting that it was the foster parents who removed the bodies of their own neglected young. In preparing his book, Gould made a sketch of cuckoo foster parents caring for a recently fledged cuckoo chick. Then, sometime around 1870, he came across a drawing by the talented bird illustrator Jemima Blackburn of a Common Cuckoo chick in the process of ejecting a Meadow Pipit chick from its nest. Having previously rejected the idea that the cuckoo chick disposed of its nest-mates, on seeing Blackburn's painting Gould was suddenly convinced and immediately set about producing a new image for his book. Nonetheless he:

> was loath to surrender his early views completely, however, and the plate he developed from Blackburn's sketch expanded the scene to include the parent pipit observing the proceedings. If such unnatural violence was to be admitted, it needed to be permitted and overseen – the pipit presumably a stand-in for a governing God.[37]

Jemima Blackburn's description of the ejection she had witnessed was published in the journal *Nature*, and she was clearly unsettled by Gould's reinterpretation of her work. She wrote to *Nature* to say: 'I have always tried in my drawings . . . to express neither more nor less than what I saw . . . none of us saw the parent pipit looking on while the young cuckoo behaved so naughtily'; and later: 'I think it right to say that I am not the authority for many of the details in [Gould's] plate'.[38] Having been exposed in this way, one wonders what Gould's subscribers thought of him behaving so 'naughtily'.

Not surprisingly, Gould was unable to stomach Darwin's revelatory explanation of the cuckoo's behaviour in terms of natural selection as set out in the *Origin*, and he makes no mention of it in his *Birds of Great Britain*.

In December 1880 the ageing Gould was visited by the Pre-Raphaelite painter John Everett Millais and his son. After being made to wait for thirty minutes, the Millaises were shown in, only to see an old man lying pathetically on a couch. On his lap was a painting of hummingbirds that he obviously wished his guests to imagine he was still working on. This pitiable attempt to bolster his status caused them to almost laugh out loud, and as they left, Millais said to his son: 'That's a fine subject; a very fine subject. I shall paint it when I have time.'

He did that painting in 1885, and the painting, entitled 'The Ornithologist or The Ruling Passion', is now considered among his greatest achievements. It shows a decrepit, bed-ridden old man surrounded by his grandchildren and their mother. In his hand Gould is holding and admiring the study skin of a red bird – a King Bird-of-paradise – with the skins

Jemima Blackburn's drawing of a young cuckoo ejecting a Meadow Pipit chick from its nest (courtesy Rob Fairley).

of various other exotic birds lying on the bed and in a box on the floor. At first sight Millais's painting looks little more than an endearing bit of sentimental Victoriana, but there is more to it than meets the eye.

Most famous for his 'Ophelia', completed in 1852, Millais was one of the founders, in 1848, of the Pre-Raphaelite Brotherhood along with six others, including William Holman

Hunt and Dante Gabriel Rossetti. Dissatisfied with what they considered the academic pretentiousness of current art, the Pre-Raphaelites decided that a new approach was needed. As the writer John Holmes has said, what they created was a 'richly imagined medievalism and an erotic symbolism [executed with] hard edged naturalism'. The Pre-Raphaelites tapped into the Victorians' obsession with the natural world, and the fashion for the close observation of nature, as Gilbert White, Leonard Jenyns and others had urged. And the Pre-Raphaelites succeeded at a level that must previously have been hard to imagine, immortalized by the photo-realism and beauty of the riverside vegetation in 'Ophelia'.[39]

Although they are not everyone's favourite group of painters, I have always loved the Pre-Raphaelites. As a sixth-former in the 1960s I was torn between going to art school or to university to study zoology. An interview at the prestigious Leeds School of Art made that decision for me. The sight of large numbers of students lying around chatting and smoking in their studios and not doing very much art ran completely counter to my Protestant work ethic. I went off to Newcastle to study zoology instead. Nevertheless, my passion for art is still deeply woven into my scientific studies of birds.

I am not sure that as a teenager I ever appreciated that the Pre-Raphaelites' art was modelled on science, even though part of their paintings' attraction for me was their accurately executed character. In the early 1800s, when scientists were busy seeking 'the truth' to help them interpret nature correctly, the Pre-Raphaelites were also seeking nature's truth in their paintings. The two groups, unlikely partners in a way, shared the same spirit of progress and discovery. The Pre-Raphaelites' knowledge of science was probably quite limited, but they liked the idea of science, of close and careful observation faithfully rendered, though they did not possess detailed knowledge of any specific scientific discipline.

The Pre-Raphaelite Brotherhood emerged at a time when science was booming, but it was also a time when science in general, and natural history in particular, were deeply entwined. The paintings, poetry and sculpture created by the Pre-Raphaelites all helped to reinforce a deep sense of natural theology; a celebration of the beauty of nature created by an all-wise God.

John Ruskin, Britain's premier Victorian-era art critic and great supporter of the Pre-Raphaelites, arranged in the 1850s for the Oxford University Museum of Zoology, then under construction, to be decorated inside and out with Pre-Raphaelite carvings: birds, beasts and botanical forms. The carvings were executed by James and John O'Shea, two Irish brothers, and their nephew Edward Whelan. The idea was to celebrate the natural world and allow visitors to obtain 'knowledge of the great material design of which the Supreme Master-Worker has made us a constituent part'.[40]

A neo-Gothic cathedral to science, and to natural theology, the Oxford Museum of Natural History, originally known as the University Museum, was completed six months after the publication of Darwin's *Origin*, and is arguably among the Pre-Raphaelites' greatest works of art. It is ironic, then, that the famous debate over God versus natural selection between the Bishop of Oxford and Thomas Henry Huxley should take place within its walls so soon after it opened in June 1860.

The *Origin* resulted in major rifts among members of the Pre-Raphaelite Brotherhood, both within science and within

Unspecified wading bird stalking among luxuriant pickerel weeds carved by John O'Shea in 1860, on the Oxford Museum of Natural History (© Oxford Museum of Natural History).

Western society as a whole. Some of the Pre-Raphaelites, including Holman Hunt and the sculptor John Lucas Tupper, stuck resolutely to natural theology, while others including Rossetti and his friends abandoned the idea of science altogether and drifted off into a world of symbolism and aestheticism.

History was soon to repeat itself, with the construction of another Gothic museum designed and decorated using Pre-Raphaelite aesthetics. Since the 1840s, when he was curator at the Hunterian Museum, the zoologist Richard Owen had been lobbying the government for a national museum to hold the country's natural-history collections. After moving to the British Museum in London in 1856, Owen renewed his efforts and in 1863 a plot was purchased in South Kensington. Work started in 1872 and Owen's 'Christian temple of nature' finally opened its doors in 1881. The close collaboration between Owen and architect Alfred Waterhouse ensured a museum more magnificent in scale and with a greater unity of design than that in Oxford. Owen was closely connected with the Pre-Raphaelites, and 'although he was ambitious, autocratic and at times a divisive figure within science, [he] was well liked by writers and artists'.

That statement by John Holmes underplays just how much Owen despised Darwin's ideas. While he did not deny evolution, he could not accept that it occurred 'by accident', through natural selection, rather than progressing along a preordained course set by God. Standing in the Natural History Museum's great hall today, one cannot help but feel the splendour of its cathedral-like construction and design. But it was too much and too late. By the time the museum opened, the idea of natural theology was all but dead, superseded by Darwinism. Worse, the Pre-Raphaelite-inspired designs within the museum's structure now seemed like a falsehood rather than the science-inspired 'truth' that they sought to promote. As Holmes notes: 'Owen's natural theological concept of the natural world was out of date before they even started to build the museum.'[41]

John Gould had known Darwin for over thirty years. But when Darwin went public with his ideas in 1859, Gould could not, or would not, embrace them. By this time Gould was a well-known public figure whose huge, beautifully illustrated bird books had made him rich and famous. He did not wish to jeopardize either his reputation or his income by being seen to support Darwin's controversial ideas.

At the time the *Origin* appeared, Gould was busy producing his multi-volume series on hummingbirds. These tiny birds were Gould's ruling passion. Like Allan Hume (whom we will meet in the next chapter) and other wealthy, obsessive bird collectors, Gould sent his own men out in search of rarer and rarer hummingbird specimens, tempting them with substantial financial rewards, including £50 (equivalent to about £4,000 today) for a Marvellous Spatuletail – which he failed to secure. Over just a few years, Gould amassed thousands of skins, nests and eggs of some 320 of the 350 known hummingbird species, and his exhibition of these in purpose-built glass cabinets in a specially constructed building at the Zoological Gardens as part of the Great Exhibition of 1851 was a spectacle few had ever seen the like of before. The tens of thousands of visitors were entranced and, inevitably perhaps, Gould's show triggered a damaging fashion for hummingbird displays and, subsequently, as accoutrements for ladies' hats and dresses: a tsunami of hummingbird slaughter. Not that Gould cared: the Great Exhibition was the best publicity possible for his five-volume *Monograph of the Trochilidae, or Family of Hummingbirds*, then in progress.[42]

Gould's hummingbird plates, designed by himself, were coloured and finished with gold-leaf and varnish to recreate the birds' iridescent plumage by the talented William Matthew Hart. While Gould's wealthy subscribers applauded his images, some ornithologists thought them inaccurate, and the art cognoscenti felt that the paintings were simply too flamboyant. Yet Gould's hummingbird volumes were among his bestsellers, and he probably cared little what critics said.

From an ornithological perspective, however, Gould's

reputation was already dead by 1859, for when the British Ornithologists' Union was being formed by Alfred Newton in 1850 Gould was pointedly *not* invited to join – a rebuff he felt very deeply.

The sparkling iridescence of the male hummingbirds' plumage was for Darwin the perfect example of sexual selection, evidence of a preference by female hummingbirds for those male traits. Gould, however, could not accept that the

Brehm's illustration of Crested Coquette Hummingbirds used by Darwin in his *Descent of Man* in preference to one of Gould's illustrations (from Brehm, 1867).

hummingbirds' exquisite beauty was utilitarian. The result was that when Darwin was looking for images of hummingbirds for his 1871 book on sexual selection, he ignored Gould's acknowledged masterpiece, and went instead to the German ornithologist Alfred Brehm, whom he knew was an avid supporter.

In 2006 Jonathan Smith, a professor of English at the University of Michigan-Dearborn, suggested that Gould's antipathy towards natural and sexual selection caused him to compose the illustrations in the *Birds of Great Britain* in an anti-Darwinian manner. That is, depicting male and female together with their nest and clutch of eggs, in morally uplifting scenes of domestic harmony, with no trace of the brutally competitive, utilitarian lens through which Darwin viewed the natural world.[43]

The ongoing march of science and its absolute requirement for specimens generated a brutishness among ornithologists. Decades before John Ray's carefully constructed and earnestly believed physico-theology was usurped by natural selection, new-found wealth among the Victorian elite fuelled a ruthless, single-minded quest for samples in the form of study skins. Blinded by the light of their own fervour, it was almost as if Victorian ornithologists failed to see the paradox that, in the process of shedding light on birds, they were casting a dark shadow over their own history. By the late 1800s our relationship with birds was about to enter a ghoulish era.

9. A Dangerous Type of Bigamy: Killing Time

The accompanying plate offers but a feeble representation of a humming bird, the beauty and elegance of which are in just accordance with the luxuriance of the glorious country it inhabits . . . The single specimen [a Blue-tailed Hummingbird, or Long-tailed Sylph], which graced Mr Bullock's Museum, . . . was for many years the only one known.

John Gould (1861)[1]

The Down House hummingbirds were the glittering tip of a deadly ornithological iceberg – a seemingly unstoppable chilly mass of enthusiastic bird killers of whom Allan Octavian Hume was the most extraordinary.

The eighth-born child, Allan was part of a prosperous colonial family. His father served as a surgeon in India, married above his station and acquired both wealth and influence, eventually becoming a Member of Parliament and the co-founder of University College London. Allan inherited his father's energy, along with his radical and liberal values.

His boyhood hobbies included nest-finding and shooting, and, like many a male scion of wealthy parents, he was groomed for a career in the colonies and, in 1849 at the age of nineteen, was posted to the Bengal Civil Service. Hunting was how male civil servants in India spent their leisure time and Hume was no different, accumulating, with the help of a 'thoroughly trained native taxidermist', a collection of bird specimens.[2] Those specimens disappeared in 1857 during the Indian Mutiny, a revolt born out of the local people's frustration with the British East India Company and the social reforms the British had imposed.

Tall, confident, with a huge walrus moustache, and simultaneously benign and domineering, Hume was a superb administrator. He was hard-working, clear-thinking and, unlike many of his superiors, deeply sympathetic to the ordinary people of India. He was contemptuous of the colonial mindset, the ineptitude of the British administration and their 'increasing tendency to ride roughshod over the wishes of the people instead of consulting and working with Indian leaders'. Hume's efforts to improve the conditions of the local people did not always endear him to his superiors.

Sometime in the 1860s Hume met Thomas Jerdon. Seventeen years his senior, Jerdon was an English physician who had arrived in India in 1836 with a passion for natural history. Recognizing the need for a written guide to Indian wildlife ('The want of a brief, but comprehensive Manual of the Natural History of India has been long felt by all interested in such inquiries . . .'), Jerdon set about producing a series of volumes on the continent's vertebrates.[3]

Starting in 1862, he produced the first of his two-volume *Birds of India*, describing no fewer than 1,008 species. Hume referred to him as his 'friend and master . . . for it was from him that I first imbibed a taste for ornithology'. And once Hume had started, there was no stopping. Throwing all his organizational powers into his new-found hobby, Hume began collecting bird specimens with renewed vigour, but also now with a much clearer sense of purpose. Enlisting the help of a huge network of contacts across India, and using his leave to visit far-flung corners of the country, he accumulated a great deal of new knowledge and many new specimens.

His first book – *My Scrap Book: Rough Notes on Indian Oology and Ornithology* (1869) – was 'a compendium not so much of his knowledge as of his ignorance', perceptively identifying what was *not* known and encouraging others with similar interests to help fill the gaps. Hume recognized that there were hundreds of British sportsmen 'in India, who could tell us facts about the nidification [nesting], habits, migrations, distribution, &c., of species of which we know little'. Known by his colleagues as 'the Pope of Ornithology', Hume 'endeared himself to all who worked for him. His enthusiasm was infectious and his knowledge of his favourite subject marvellous.'

His energy was boundless, but he was frustrated by the constraints associated with publishing in the British Ornithologists' Union journal, *The Ibis*. As a result, he decided, at his own expense, to start his own journal, *Stray Feathers*, in 1872, with Hume reading, reviewing and editing his colleagues' manuscripts as well as writing many articles himself. His aim, as always, was the acquisition of new knowledge, with the ultimate goal of producing his own definitive account of the birds of India. And all this while working full time as secretary to the new Department of Revenue, Agriculture and Commerce; the highest point in his Indian Civil Service career.

In 1871 Hume acquired a substantial house, Rothney Castle, as his new summer home near Shimla, within sight of the snow-capped Himalayas. And, it was here, 2,100m (6,889 feet) above sea level, that Hume established the ornithological museum that was eventually to house his collection of 18,500 eggs and 80,000 study skins. This vast array of specimens enabled Hume to plot the geographic distribution of individual species, to recognize different subspecies (or races) and to discover a host of species new to science. No fewer than seven birds now bear his name, including Hume's Wheatear, Hume's Leaf Warbler, Hume's Short-toed Lark and Hume's Treecreeper. His extraordinary energy was focused on just one thing: his planned *Birds of the British Indian Empire.*

Hume was just one among many amateur ornithologists across much of the world amassing bird skins and eggs. For 200 years, starting in the mid-1700s, this was how ornithology was done. Without specimens, the scientific study of birds was impossible, and hundreds of would-be ornithologists began collecting. It was an endeavour made possible by the perfection of preservation techniques.

The conservation of acquired specimens was fundamental to the study of birds; without imperishable study skins there was no proof of their existence, no reference point for comparison and only limited opportunities for identification. Pictorial representations were only as good as the artists who created them, and while sometimes very simple illustrations, such as those in the El Tajo cave, would allow identification, they were hardly definitive. Even the extraordinarily detailed bird paintings of Renaissance artists like Albrecht Dürer and Pisanello were less useful than an actual specimen.

As we've seen, the earliest form of animal (and human) preservation was desiccation and mummification – and very effective it was too. The curing of animal skins as clothing predates mummification, and although we tend to think of our ancient ancestors wearing mammal skins, bird skins were also used. In the 1500s, and no doubt long before, the indigenous people of the Arctic regions made caps from the entire skins of birds like divers (loons), whose soft, well-insulated and waterproof plumage was perfect for such a purpose. After having all flesh and fat removed, these bird skins may simply have been dried in the sun to preserve them.

Frederick II dried the wings of cranes, 'hung in the chimney until the moisture in the flesh has completely evaporated', to use as fake birds in the training of falcons. Similarly, the skins of birds of paradise that first arrived in Europe from the Moluccas in the 1500s had been dried in the sun or over a fire. The colours, iridescence and texture of the feathers were unlike anything anyone had seen before. The lack of feet, removed by those who skinned them, was naively assumed to be their natural state, forcing them to fly continuously 'in paradise'.[4]

Numerous works of art, including Jan Breughel's 'Allegory of Taste' (1618), now in the Prado, show fully feathered birds dished up at sumptuous feasts. These were skins, preserved, presumably by drying, and used as pie covers. The famous French chef Guillaume Tirel, in the service of both Charles V and VI in the 1300s, served 'roast swan and peacock, redressed after cooking in their entire plumage, complete with gilded bills ("endored") and an appropriate miniature landscape sculpted in sugar'.[5]

It was only as cabinets of curiosities started to become popular in the mid-1500s – and collectors became increasingly anxious to safeguard the exotic specimens brought back by voyagers – that preservation techniques began to improve. The ornithologist Pierre Belon, writing in 1555, describes how one should remove a bird's internal organs, including its brain, add salt to the inside of the skin and then hang it up to dry. Later, in the 1600s, the German bird-catcher and writer Johann Conrad Aitinger suggested using ash, alum and sulphur to help preserve the skin, which was then pulled over a straw 'corpus', using wires and skewers to hold the bird in a lifelike posture. These specimens still needed to be dried over a fire 'every quarter of a year' to protect them from the scourge of moth and beetle larvae that continue to be the occasional bane of museum curators to this day. Ironically, the preservative concoction developed by Aitinger damaged the skin, so was less effective than he hoped.

A woodcut of a Eurasian Coot, based on a badly stuffed specimen, from Conrad Gessner's bird book of 1555.

We also know that many of the illustrations in the first bird books, such as that of Conrad Gessner in 1555, were drawn from crudely mounted, stuffed specimens, hence their distinctive 'wooden' or overstuffed attitude.[6]

The turning point in the preservation of bird skins was the discovery of arsenic soap in the 1680s. 'Take white arsenicum and twice as much water of copper, grind it all nicely on a grindstone and add strong brandy to thin it . . . Then, when you skin a bird the skin should be painted on the inside with the mixture.' This will 'keep these birds safe from cockroaches and worms, so that they remain beautiful even when they are old and not a single feather falls out'.[7]

Arsenic soap smeared on the inner surface of a bird specimen meant that study skins were no longer vulnerable to the jaws of voracious insect larvae. Better guns, and better methods of travel, further facilitated the new craze for collecting. And it was justified as science. Creating what were called 'study skins' was distinct from the preparation of birds mounted in lifelike positions for public display. Both study skins and mounted specimens require proper preservation, but mounted specimens also require an internal armature and special skills to recreate a realistic posture. Study skins, on the other hand, are eyeless, functional and compact, to save on storage space, while retaining their main features.

In the late 1960s I was lucky to be able to see a master taxidermist, Reg Wagstaffe, prepare a study skin of a female Peregrine Falcon. The bird had been found dead and handed in to Monks Wood Experimental Station, which at that time was one of the most important sites for ecological research in the UK. This was also where much of the research on the effects on wildlife of pesticides such as DDT was being undertaken. I was visiting Monks Wood for a couple of weeks because of my interest in Grey Herons, which, like Peregrines, are susceptible to pesticide poisoning. At the time, I had no idea who Reg was. He simply seemed to be a kindly old chap who talked me through his masterful transformation of a sad, pesticide-poisoned predator into a beautiful biological specimen. The modestly triumphant finale involved Reg picking up the bird and gently wrapping it in soft woollen thread, telling me that this was how it would maintain its streamlined shape – as though in a stoop – until it had dried. It was as though he was wrapping a mummy.[8]

It wasn't quite the end. The last thing Reg did that afternoon was to write out a label, an oblong piece of card, specifying the bird's details: its identification (*Falco peregrinus*), its sex (female), location and date of death. He then tied the label to one of the bird's legs.

Labelling is essential. Without a label, a study skin, or any other biological specimen is of limited value. Eggs make the point. Natural-history museums contain huge numbers of birds' eggs, many of them unlabelled because their owner or collector was reluctant to spoil the egg's beauty by writing on it. Instead, collectors sometimes pushed a piece of paper bearing a reference number (linking it to a catalogue) into the hole through which the egg's contents had been removed. Sadly, this very impermanent form of labelling failed over time as the paper dried up and fell out and as catalogues became separated from the eggs they listed. A minority of egg collectors wrote the details on the egg itself, but this was rarely sufficient.

Nowhere is the value of labels better illustrated than by a scandal perpetrated by someone once considered Britain's most important ornithologist: Richard Meinertzhagen. A wealthy aristocrat, whose wife died in a suspicious accident

The egg of a Brünnich's Guillemot collected on Novaya Zemlya in the early 1900s and now in Tromsø Museum (photo: Tim Birkhead).

involving a revolver while he and she were out together, Meinertzhagen was a soldier, spy, ornithologist and pathological liar. He was an arrogant, archetypal member of the British Empire, who, in complete contrast to Hume, could not imagine India ever governing itself, harbouring a special hatred for Mahatma Ghandi.

Like Hume and other Victorian ornithologists, Meinertzhagen's focus was on identification and establishing the geographic distribution of different bird species. Considered 'the last of the great British bird collectors', Meinertzhagen collected birds across Eurasia between the 1920s and the 1940s, and was a frequent visitor to the Natural History Museum in London to make comparisons between his study skins and the museum's specimens. For reasons best known to himself, Meinertzhagen had an unquenchable craving for recognition and respect, and used falsified 'range extensions' – *new* records for wherever – as one way to achieve that. He stole museum specimens and relabelled them with fabricated data, and presented them as his own. Sacrilege. When, in the 1990s, long after his death, Meinertzhagen's fraudulent activities were discovered by Pamela Rasmussen and Alan Knox, such was the despair at ever sorting out the mess he had created that the chief curator suggested Meinertzhagen's entire collection of 20,000 study skins should be burnt. Luckily, that did not happen, and through some ingenious if laborious forensics, researchers were able to rectify much of the mayhem Meinertzhagen's mislabelling had created.[9]

In 1998 two literary ornithologists, Barbara and Richard Mearns, published an all-embracing account of the bird collectors of old, listing the hundreds of individuals, mostly Europeans and North Americans, and mostly men, who collected birds and their eggs across different parts of the world.[10] Among the most acquisitive and successful was Walter Rothschild. In 1875, at the tender age of seven, Walter announced to his parents that he would start a museum. His 'precocious zoological genius and the fabulous wealth of the Rothschilds . . . meant that . . . his childhood ambition came to abundant fruition'. Once his father accepted that Walter would not follow in his banking footsteps, he indulged his son – to the tune of giving him the family estate in Tring, Hertfordshire, and £1 million (equivalent to over £100 million today) for his ever-expanding museum. This allowed Walter to employ hundreds of collectors across the world to seek out and secure rare birds. To help run his ever-expanding museum, Walter took on Ernst Hartert as curator in 1892, freeing himself to write up the results of his studies. One of the most striking of these is a monograph, beautifully illustrated by John Gerrard Keulemans, a leading bird artist of the day, on the cassowary (native to New Guinea and northern Australia), of which Rothschild had sixty-four *live* specimens at Tring. Overall, Rothschild, Hartert and a second curator, employed in 1893, produced a total of 1,200 books and articles, and described 5,000 new species based on Walter's collection of 300,000 bird skins and 200,000 eggs.[11]

But then, disaster. Walter never married, but had at least three well-to-do mistresses, the identity of only two of whom are known. Both blackmailed him to the tune of £10,000 a year, but it was the third who finished him off. Known only to be a married 'peeress', she (together with her husband) threatened to make Walter's affair with her public, and for forty long years screwed money out of Walter so relentlessly that he was eventually forced to sell his huge collection of bird skins. The beneficiary was the American Museum of Natural History in New York, whose curators felt they had struck an extraordinary bargain at $225,000. And Walter? He died a few years later, in 1937.[12]

It was inevitable, given the era, that most collectors were men, but many of them were supported by a spouse, including Frank Chapman, who for over fifty years worked for the American Museum of Natural History in New York. He said this about his wife Fannie:

> When a man wedded to his profession takes a mortal wife he commits a very dangerous type of bigamy. If the two spouses do not agree there arises a three-cornered conflict

> to determine which one of them will be widowed. If they are in harmony, a man may indeed consider himself twice blest. I was among the fortunate ones.[13]

Among the handful of female collectors, perhaps the most outstanding was Emilie Snethlage, one of the first women in Germany to obtain a PhD. Clever, tough and independent, she subsequently spent much of her life in South America exploring remote regions of the Amazon, living with local tribes, collecting thousands of bird specimens and describing many new species – including one known now as Snethlage's Tody-tyrant. She died from heart failure aged just sixty-one, scotching her plans to spend her last years in Germany writing up her work. When Helmut Sick published his *Birds in Brazil* in 1993, he dedicated it to her.

How did people like Allan Hume, Walter Rothschild and Emilie Snethlage justify their level of killing?

Hunting and collecting have always been obsessions among the wealthy, whether it was the pharaohs fowling in the marshes and filling their tombs with artefacts, Inca chiefs with their menageries, or Europeans like Ole Worm and Francis Willughby cramming their cabinets with curiosities. The obsession with bird collecting in the 1800s and 1900s was a continuation of that, but much more widespread, because by this date a higher proportion of people in Europe had the wealth and time to collect. Acquisition and accumulation were – and still are – deep-rooted cravings for status.

Chatsworth House in Derbyshire, close to where I live, is owned by the Cavendish family and is a popular tourist attraction, mainly, I think, because the house is so stuffed with collectibles as to defy belief. The treasures include works of art, furniture and books. I once used the library there only to discover that the *two* full sets of Buffon's *Natural History* (thirty-six volumes each) had probably never been opened since the day they were bought in the early 1800s. Through the 1700s, easier travel and better firearms encouraged the collecting of wildlife, but it was science that gave collecting its biggest boost.

Starting with curiosities collected by savants in the 1500s, the Scientific Revolution in the mid-1600s provided the impetus and legitimacy for a two-century-long collecting spree. Because God had given man dominion over nature – the freedom and wherewithal to exploit the natural world as he saw fit – religion and science were inextricably entwined.

For many bird and egg collectors, the primary motivation was the thrill of the chase: the hunt itself. 'A curlew was the first bird I shot, and though it is close on forty years ago, that exciting stalk is never to be forgotten,' wrote Colonel C. T. Bingham in 1895.[14]

I have asked old egg collectors about their motivation, and they say the same. Looking at particular 'sets' of rare eggs in their collection allows them to relive the finding of the nest and then getting to take the eggs, little different from a twitcher seeing a new bird for the first time. Collecting was predominantly a manly pursuit, requiring excellent field craft, stamina and endurance. It was often risky and exciting: an excuse for men to behave like boys.

Willoughby Verner, whom we met in the first chapter, is a prime example, illustrating his books with scenes of derring-do, including of himself taking the eggs from a Lammergeier nest on a vertiginous cliff. Competition was another dimension to all this, driving men to accumulate the biggest or best collections, and to travel to increasingly remote and dangerous parts of the world in the hope of hitting the jackpot and discovering new species. Of course, most of these men also *liked* birds: they enjoyed seeing them, handling them, skinning and sometimes, dissecting them. The curious mix of sentiments is captured by Edgar Layard, who, on killing his first Orange Fruit Dove, a stunningly beautiful bird, while in Fiji in the 1870s, wrote: 'Picture to yourselves our delight as we handled the brilliant Orangebird, with the sunlight gleaming through its golden wings, and lighting up the emerald green of its bill and feet, and of the cere round the yellow-buff eye, now closing, alas! in death.'[15]

Above everything else was the idea that there was no other way to study birds.

The world's museums currently contain around ten million bird skins. Over 99 per cent of the 10–11,000 known bird species are represented in these collections. No other animal group, except mammals, of which there are far fewer species (just 5,400), is so well represented in museum collections, and as such they represent the most remarkable scientific resource.

Ten million birds in museums? I can feel you squirming uncomfortably, but wait. Domestic cats kill this number of birds *every day*, and similar numbers are destroyed by hunters, habitat loss and climate change. And with respect to chickens . . . each day sees the death of over 100 million individuals.[16]

A chicken meal is digested in around twenty-four hours. In

contrast, museum specimens serve us for much longer. A major criticism levelled at the collectors of the past – and this is true of birds collected for their skins, but also eggs – is that the collectors' focus on rare species helped to drive those very species towards extinction. A well-known case in Britain is the Red-backed Shrike, whose attractive and variable eggs were relentlessly collected, most notably in the 1920s by Edgar Chance, who is better known for his studies of the Common Cuckoo, whose eggs he also avidly collected.

Chance was a wealthy, complex individual, described as 'prickly, dogmatic, grasping, paranoically suspicious of rival collectors, but also stimulating, exuberant, enthusiastic and compelling'. Britain's Red-backed Shrike population declined throughout the twentieth century, and by 1988 was reduced to a single pair. No shrikes bred the following year and there seems to be little doubt that the species' rarity encouraged egg collecting and hastened its extinction in England.[17]

The Guadalupe Caracara, known to exist only on Guadalupe Island off Baja California, was driven to extinction in 1900 by the supremo professional bird collector, Rollo Beck:

> Although I had no idea of it at the time it seems probable to me that I secured the last of the Guadalupe Caracaras . . . on the afternoon of December 1, 1900. Of the 11 birds that flew toward me 9 were secured. The other two were shot at but got away. The 11 birds were all that were seen, but judging by their tameness . . . I assumed at the time that they must be abundant.[18]

Beck sent the nine skins to Walter Rothschild, and that species was never again seen in the wild.

Another species whose last individuals were taken by collectors was the magnificent Ivory-billed Woodpecker, native to the swampy hardwood forests of the south-eastern United States. The relentless logging that characterized the 'development' of much of North America, especially after the Civil War, drastically reduced the woodpecker's habitat and population. The last truly authenticated sighting was in 1944. Since then, and most recently in 2004, there have been claims that the species still exists. I was able to join one of the groups that believed this to be true, and although canoeing through the flooded forests of the Choctawhatchee River in Alabama was wonderfully surreal (reminding me of the film *Deliverance*), I had little expectation of seeing an Ivory-bill.

For most of his time in India, Allan Hume served and thrived under a succession of liberal viceroys. The appointment of the conservative Lord Lytton in 1876, however, brought about a change. Frustrated by Hume's relentless and public criticism of both himself and the system, Lytton decided in 1879 to abolish Hume's department and demote him.

Humiliated, Hume remained in post, for he needed his salary to cover the costs of his book on Indian game birds then in production. Once that was published, Hume retired – he was fifty-two – and settled down at Shimla to organize his vast collection of notes accumulated over the previous fifteen years in preparation for producing his *Birds of the British Indian Empire*.

During the winter of 1882–3, while Shimla was under snow, Hume and his wife took an extended break on the plains, where the weather was milder. On their return, Hume found to his horror that his precious notes had been stolen by a disgruntled servant, who had taken them, together with all 6,000 pages of his museum catalogue, and sold them as waste paper in the local bazaar.

Hume was crushed. 'The dream of his life had . . . vanished. No hope was left that he would ever reap the reward he had laboured for many years to gain. The great book could never be completed.'[19]

It is hard to imagine just how much of a blow this was to Hume. Our experience of losing the contents of a computer hard drive doesn't come close. As Hume wrote to his colleague Richard Bowdler Sharpe: 'I have no heart to undertake the re-writing, for ornithology has no longer the interest for me that it once possessed.' So ended Hume's ornithological career. As his friend and co-author Colonel Charles Marshall wrote: 'Few knew how deeply he felt having to come to this decision, for he said but little.'[20]

Devastated by demotion, loss of status and focus, Hume turned to the new pseudo-religion of Theosophy that, like the Indian faiths, claimed to respect all forms of life. Founded in New York in 1875, but struggling to establish itself there, the movement's founders, Helena Blavatsky and Henry Olcott, moved to India in 1879, which is where Hume first encountered them. He liked their humanitarian beliefs of brotherhood and social improvement (though whether he agreed with their belief in chastity within marriage is unknown). Hume joined the group and gave up drinking alcohol, eating meat and killing birds. Earlier in his life he had had qualms about killing, but hardened himself in the name of science:

> The worst of ornithology is having to kill birds like these [Siberian Cranes] . . . I do not know how it is; but I have often wished that I could be quite sure that the wholesale

> murder of these and similar innocent animals merely for scientific purposes, and not for food, was quite right. Intellectually, I have no doubt on the subject; but somehow, when a poor victim is painfully gasping out its harmless life before me, my heart seems to tell me a somewhat different tale.[21]

Hume's Theosophy persuaded him, in the late 1870s, to stop killing birds for science. A century later the morality of collecting birds for science raised its head again, over a bird Hume had discovered in 1873.

On 4 February 1986, while in Eilat, Israel, an experienced birder came across a tiny brown bird he was unable to identify. Such was the craving to name the mystery bird, it was trapped and flown (first class) to Tel Aviv so it could be observed at close range in the Zoological Gardens, where it died a few days later. Duly transformed into a study skin and given a place in the university museum at Tel Aviv, the original finder later took the bird with him to the Natural History Museum in Tring. Comparing it with some of

Hume's Short-toed Lark, painted by Per Alström from a combination of field sketches in China and India and museum studies, for a paper on the identification of this species.

Hume's specimens, he was able to identify the mystery bird as a species new to Israel, Hume's Short-toed Lark, and well outside its normal range of Nepal, China and the Ladakh region of India.[22]

When this new record was reported in the pages of *British Birds* in 1990 there was an outcry, with readers appalled that the bird's welfare had been sacrificed simply to acquire a Western Palearctic 'tick'. Another respondent countered by saying that the reaction was 'oversentimental and unscientific', reminding readers that the natural death rate of small birds is such that 'whether a particular bird survives or succumbs is of no significance' – the 'individual versus the population' dilemma once more.[23]

A similar case involved an unidentified shrike seen first in the grounds of a hospital in the town of Bulo Burti, Somalia, in 1988. The bird's unusual appearance suggested that it was an entirely new species of Boubou Shrike. It was duly captured and, because of political unrest in Somalia, flown to Germany, where it was maintained in an aviary. Fourteen months later it was flown back to Somalia (where the political turmoil was even worse) and, after being photographed and measured, and a blood sample taken for DNA analysis, it was eventually released in 1990, contrary to usual practice. The bird was indeed new to science. It was given the scientific name *Laniarius* (shrike) *liberatus* to celebrate the fact that, instead of ending up as a study skin in a museum drawer, it had been released. The Bulo Burti Shrike made headlines around the world as the first new bird species to be described from a live specimen. Later, using the preserved blood sample, a subsequent molecular analysis revealed that the bird was not new after all, but simply a colour variant of a well-known species the Coastal Boubou *Laniarius nigerrimus*.

The issue of detached collectors is far from dead. In April 2017 Francisco Sornoza-Molina of Ecuador's Instituto Nacional de Biodiversidad was working in the *páramos* (moorland) habitat of the western Andes in Ecuador when he noticed an unusual hummingbird. Sornoza-Molina and his colleagues subsequently found other individuals of the same unidentifiable bird and, thinking it might be a new species, trapped and collected (that is, killed) seven individuals. By comparing the study skins of these specimens with those of similar hummingbird species in museum collections, it was obvious that it was a new and previously undescribed species. They named it the Blue-throated Hillstar and their account, published in *The Auk* in 2018, included a comment that the bird was rare and probably critically endangered.[24]

Just as with Hume's Short-toed Lark and the Bulo Burti Shrike, the paper triggered a controversy, in this case on a birders' blog site, where it was asserted that:

> The days when it is necessary to kill extremely rare animals in order to describe them scientifically have long gone. In the case in hand, excellent photographs were taken, and DNA was collected and used to confirm the identification . . .
>
> It is perhaps even more reprehensible that a respected journal of a respected ornithological society [the American Ornithological Society] should consider it suitable for publication.

Collecting for museums was an important part of the scientific past of ornithology, and there is still a case to be made for preserving specimens in museums. However, there can be no justification for killing a new species of

> bird, when you consider it to be critically endangered. No doubt when IUCN/Birdlife International publish the next Red List, they will include this species, and among the reasons to be given for its endangerment, will presumably be 'collecting for museums'.[25]

Like-minded others chipped in: 'utterly appalling . . . it cannot be justified on any level'; and, 'removing seven specimens from the gene pool of a critically endangered species outweighs what marginal gain to science there might be in collecting them'.

Another contributor, however, felt that this was all 'inflammatory nonsense' and that 'the collection of specimens for deposition in a museum collection and subsequent study is still the benchmark for the description of new species – and for very good reasons'.

The tirade continued, with most respondents opposing the collecting. Then came a lengthy, sober response from Guy Kirwan, an experienced bird taxonomist, whose first point was that 'non-taxonomists do not know how to practise taxonomy properly. It too is a science.' He then explained the process and how it is regulated. Drawing attention to the almost unimaginable numbers of birds inadvertently killed by human activity – in windows and by traffic – compared with those collected for museums, he then suggested that for those who have an ethical problem with museum collecting, it is 'beholden upon you to go as far as you can to minimize your personal contribution to the many other (far worse) impacts on birds and their habitats that can be laid at the door of *Homo sapiens*'.

One of the authors of the original article, Richard Porter, had previously made a point not considered in this hummingbird exchange, which was that he opposed any collecting on his expeditions:

> Because much (in fact all) of my research/survey/conservation work involves working with natives of the country concerned I believe that collecting/killing sends out totally the wrong message. Very often meeting me or one of my Western colleagues is their first experience of 'conservationists'. I want to convey a view that ALL wildlife should be respected, and that I, as a conservationist, care for and respect ALL wildlife. It's really as simple as that.[26]

Most of the pros and cons of collecting bird specimens for museums had been aired almost thirty years previously by James Remsen, curator of birds at Louisiana State University, in a lengthy, carefully considered overview of this contentious issue. He points out that those opposed to collecting have difficulty 'identifying the critical units of concern, namely habitat and the population it supports, from their concerns over the welfare of individual animals'. He concludes his article by saying, 'Museum scientists and conservationists are natural allies who should work together to achieve mutual goals rather than clash over minor, counterproductive issues such as if and how many specimens should be collected.'[27]

Remsen's account is a defence of collecting, as we might expect from a museum curator working in North America. The culture of collecting continues there, even to the extent that undergraduates on ornithology courses are taught how to make study skins. Among European scientists, collecting is almost non-existent, as are 'ornithology' courses at university level. This is a difference that has always intrigued me. Zoology degree courses at European universities often contain lots of information about birds, but it is usually taught within the broader framework of ecology, evolution or animal behaviour, rather than as a taxon-specific subject, as occurs in many North American universities.

The attitudes of American and European ornithologists towards collecting differ dramatically – a difference that may reflect national characters: American enthusiasm versus European reserve; guns versus no guns. I asked a senior ornithologist closely and impartially involved in the collecting debate how this difference arose and why it persists. Wealth, or, more politely, funds, he suggested, are at the heart of it. For Europeans, the cost of two world wars together with the loss of imperial territories pretty well put an end to overseas collecting, and Walter Rothschild's fall from grace was almost the final nail in the European collecting coffin.

American museums continue to be supported by wealthy benefactors, allowing them to maintain their collecting tradition. As the anti-collecting sentiment gained momentum in the mid-twentieth century, the Americans had much more to lose than the Europeans. Better budgets have also allowed American museums to develop molecular laboratories, such that they are often ahead of the game when it comes to identifying new species and the relationships between birds. And that means the emphasis in America is more museum- than field-based; they have also taught South American ornithologists to collect, at the expense of observing birds.

For those who collected birds in the 1700s and 1800s, their justification was advancing science: documenting the world's birds, describing the birds and their geographic distribution, all with the aim of building the foundations of ornithology.

Did they recognize this at the time? Probably not. Only with the benefit of hindsight are we able to see how scientific disciplines develop. The 'bird-collecting era' was typical of the descriptive phase of many disciplines.

Since the day of the big collectors, the additional value of museum collections has become more and more apparent. They represent vast inventories of biodiversity: catalogues of the natural world. The information we have on birds is more complete than for any other taxa. Whereas collectors like Hume, Rothschild and Snethlage used morphological features to deduce the taxonomic affinities of particular species, the surge in new methodologies, notably in molecular biology, has created new ways to elucidate taxonomic relationships, and often use DNA from museum specimens. Rothschild, even with all his money, could not have anticipated the study of genomes, and the same will undoubtedly be true in the future, with new techniques allowing us to address new questions.

Museums are important for what they can tell us of the life history, ecology, morphology and conservation of birds, but they are also important for studies of human health and our understanding of infectious disease. Many of the microorganisms that cause diseases like bird-flu (H1N1), psittacosis (Chapter 7), SARS and probably Covid-19 are 'zoonotic', meaning that they can be passed between people and nonhuman animals. Reference collections in museums allowed researchers to reconstruct the influenza virus that killed around fifty million people worldwide in 1918. The same could be true of Covid-19. The authors of one study said, 'Collections can provide short-cuts to public health responders looking for the origins and distribution of disease, but only if the collections are accessible and well documented.'

This stark warning is especially poignant, given that over the last few decades museums in Britain (and elsewhere) have been grossly underfunded and reduced to one or two staff, who can do little other than keep their heads above water.[28]

One of the most puzzling aspects of museum collections is the difference in the way the public perceives bird study skins and birds' eggs. There seems to be much less disquiet about the ten million study skins lying in museum drawers than there is about collections of eggs. Putting aside for now the argument that, in both cases, the number of museum specimens is only a tiny fraction of what is available, killing adult birds is potentially more damaging to a population than taking eggs. This is because many birds have evolved to cope with high levels of nest predation and will often re-lay right away if a clutch is taken. This is particularly true of passerine birds like European Robins, Common Blackbirds and warblers, whose lifestyle is to breed fast, die young.

The Zebra Finch is typically ready to breed at three months old, produces two, three or four broods in rapid succession and then dies within the year. For such birds the loss of some clutches is normal and their populations still have huge potential to grow. At the other extreme, many large non-passerines such as albatrosses, eagles (and the Great Auk), lay only a single egg each season and do not replace it if it is lost. Their strategy is long and slow: low reproductive output coupled with extended parental care and a long life. Some albatrosses can reach forty, fifty or occasionally sixty years of age and lay a single egg only every second year.

Eggs are emotive, stemming perhaps from a deep-seated

parental instinct. The emotional response to the taking of wild birds' eggs means that they are rarely on display in museums for fear of causing offence. A negative response to eggs is hardly innate, and some individuals are actually obsessed by birds' eggs. But our perception of eggs and egg collecting probably harks back to the same kind of God-fearing emotions many Victorians had towards birds. Emotion is perhaps the most powerful tool we have for protecting birds, but it can be riddled with contradictions.

The reduction in egg collecting since its peak in the 1920s and 30s has been dramatic thanks to the activities of non-government agencies like the Royal Society for the Protection of Birds in the UK and the National Audubon Society in North America. Collecting eggs is now illegal and socially unacceptable. Anyone found guilty of collecting eggs is likely to be prosecuted and publicly ostracized. However, it is probably true that, in the past, despite its widespread nature, the casual collecting of common birds' eggs by boys (and it was largely boys rather than girls) had no effect on bird populations and, as admitted by several famous naturalists, was an important part of their apprenticeship. The vast majority of egg collectors grew out of it during their teenage years, but some didn't and used the replacement-clutch argument to defend their activities. This was disingenuous, as most collectors were not bothered about the eggs of common songbirds, instead wanting those of rare, unusual, difficult-to-get species, and this often meant raptors: Golden Eagles, Peregrines and Western Ospreys. Their lust was fuelled by the fact that these species also had beautifully marked eggs.

The battle between collectors and the authorities is epitomized by the stealing of Scotland's Loch Garten osprey eggs in 1958. Having become extinct in Britain in 1916, one pair of Western Ospreys returned to Scotland to breed in 1954 and drew egg collectors like sharks to blood. The eggs were stolen. Increased security foiled subsequent attempts, and then in an act of frustrated revenge, egg collectors felled the nest tree in 1986.[29]

For collectors like these, their boyhood hobby had metamorphosed into an adult obsession. Psychologists exploring the motivation behind extreme collecting have identified a number of traits that predispose men (again, mainly men), to become pathological collectors. One of these is insecurity, usually through the loss of a parent as a child, and the subsequent need to compensate as an adult by accumulating 'stuff'. One infamous collector that I knew of, albeit indirectly, was John Robjent, who collected eggs from the birds breeding on his tobacco farm in Zambia, and whose meticulously curated collection of Cuckoo-finch and Tawny-flanked Prinia eggs was later used to investigate the evolution of brood parasitism in birds. Robjent was a classic case of the obsessive collector whose mother had died when he was a boy.[30]

The clearest example I have seen where collecting becomes pathological was at an exhibition in London in 2018. Organized by a father and son, Peter and Andy Holden, the centrepiece of this exhibition was a replica collection of some 7,000 eggs confiscated from Richard Pearson, a notorious collector. Other exhibits included some nests, a giant bowerbird's bower (not a nest), and some video clips of 'interviews' with two convicted egg collectors. Having studied and written about the biology of birds' eggs, I was asked if I would lead a public tour of the exhibition. Not wanting to go in cold, I went the week before to suss it out. I was shocked, first by the amount of biological misinformation and second by the self-righteous evangelical zeal with which the exhibition organizers vilified egg collectors. Watching the videos of the two egg collectors ranting against their persecutors, it was startlingly obvious, to me at least, that they were more in need of help than harassment. Yet, as the video played, it was clear that the audience were fully on board with the exhibition organizers, reminding me how easy it is to manipulate people's attitudes.[31]

As my tour with the public reached the exhibition's finale – the room containing replicas of the 7,000 eggs as they had been found under Pearson's bed in 2006 – I noticed the exhibition organizers in the audience. As I reported their statement, that after Pearson had been convicted the authorities had destroyed his collection – presumably in the hope that this would deter other would-be collectors – an authoritative voice from the audience called out to say that this was untrue. This was none other than the curator of eggs from the Natural History Museum in Tring, Dr Douglas Russell. The eggs had *not* been destroyed; they were now in the Natural History Museum.

I am not defending egg collecting. In Britain and elsewhere, it is against the law and it is wrong, but using misinformation to manipulate public opinion, together with excoriation of individuals clearly in need of psychological help, is also wrong.

Let me be clear about another thing. Destroying the collections of convicted egg collectors is also counterproductive. Much better that those collections be curated and cared for in a museum with the potential that they can be put to good use by researchers.

Without egg collections we would not have known about the hugely negative effects of pesticides on eggshell thickness that contributed to the catastrophic decline of birds such as Peregrines and Bald Eagles in the 1950s and 1960s. Nor would we have known about the negative effect of acid rain – reflected by the reduction in eggshell quality in certain bird populations. Since egg collecting became illegal, the only eggs now coming into museums are those confiscated from convicted collectors and they should not be wasted. With climate change upon us, it is not difficult to imagine that museum specimens may some day help us better understand its effects.

Visiting the Natural History Museum in Tring for my research in 2019, I noticed a gap in a tray of Pin-tailed Sandgrouse eggs where an egg had once lain. I summoned the curator, who, rolling his eyes, said with a sigh: 'Ah yes, one of Shorthouse's thefts and one of the few instances where he failed to replace the stolen egg with a substitute.'

Mervyn Shorthouse turned up at Tring in 1975 in a wheelchair, claiming to have had an electrical accident, and saying that his only solace was looking at birds' eggs. Could he volunteer to help with the collection? As the curator Michael Walters later wrote: 'The Museum took pity on him, and he was permitted to visit over a period of about five years. He was a nuisance, but we did not wish to be accused of prejudice against the disabled.' When, in 1976, a researcher noticed a clutch of Great Bustard eggs he had looked at a month previously was missing, the police were called. Dutifully, the young constable explored the different possible explanations for the missing eggs and asked Walters if he was sure the eggs hadn't just hatched. Shorthouse was duly identified as the thief, and when the police visited his home they discovered the *ten thousand* eggs that he had smuggled out of the museum tucked into the ladies' tights he wore under his trousers. He got two years and the eggs went back to the museum.[32]

Compare this with the feather thief Edwin Rist, who posed as another innocent visitor to the Tring museum, but then returned one night in June 2009 to steal 299 trogon and quetzal study skins, whose feathers he sold (often individually) to those who make artificial fishing flies. After pleading guilty, Rist received a twelve-month jail sentence, suspended in view of his alleged Asperger's and his career as a musician. Many of the stolen skins were eventually recovered, but by removing their labels Rist destroyed most of their scientific value. Letting Rist off so lightly suggests that the judge had little appreciation of the value of museum specimens. Perhaps if the theft had involved eggs rather than skins, Rist might not have been so lucky.[33]

It was the theft of his ornithological papers in 1883 that brought Allan Hume's ornithological career to a juddering halt. Almost immediately, he wrote to Richard Bowdler Sharpe, the curator of birds at the British Museum, suggesting that they might take his collection.

Hume specified conditions, including that Sharpe should come to India for eight months to catalogue and pack the specimens, and that the museum should increase Sharpe's salary and rank accordingly. The museum administrators weren't happy with this and suggested an alternative that Hume didn't like. Discussion went back and forth, and while negotiations were ongoing, heavy rains caused the collapse of the hillside on which Hume's museum was built, with the eventual loss of 20,000 specimens. The British Museum eventually acquiesced and sent Sharpe to Rothney in May 1885. Over the next three months he packed up and shipped back to Britain Hume's remaining collection of 63,000 bird skins, 500 nests and 18,500 eggs.[34]

Hume was an ebullient, desperately driven enthusiast, impatient of those – senior or junior – unable to keep up with him or meet his high standards. It was his uncompromising behaviour, no doubt, that encouraged Hume's servant to plan and execute his master's ornithological downfall with such excruciating efficiency. His ornithological lights extinguished, Hume 'devoted himself unswervingly to politics; and like his father, his radicalism increased with age', railing against 'the poverty that British maladministration imposed on the Indian people'. On moving back to Britain in 1894, he reinvented himself as a botanist, accumulating a collection of British plants that became the envy of the British Museum.[35] At the pinnacle of his career Allan Octavian Hume was the archetypal Victorian ornithologist. By the time he died, aged eighty-three, in July 1912, a completely new way of being an ornithologist was emerging.

10. Watching Birds: And Seeing the Light

> By going among the birds, watching them closely, comparing them carefully, and writing down, while in the field, all the characteristics of every new bird seen . . . you will come easily and naturally to know the birds that are living about you.
>
> Florence Merriam (1889)

Like other nineteenth-century ornithologists, Edmund Selous killed birds to study them. But in June 1898, when he was just forty, he had an epiphany while watching a pair of European Nightjars.[1]

Magical and enigmatic, the nightjar's perfectly patterned plumage provides exquisite camouflage while it is on the ground, as Selous discovered as he stared out from his hide. He knew there was a bird incubating in front of him, but it took over an hour before 'I finally became convinced it was the bird and not a piece of fir-bark at which I was looking; and this though I knew the eggs to be there'. Thrilled by what he had seen, he wrote:

> I must confess that I once belonged to this great, poor army of killers, though, happily, a bad shot, a most fatigable collector, and a poor half-hearted bungler, generally. But now that I have watched birds closely, the killing of them seems to me as something monstrous and horrible.

He continues:

> The pleasure that belongs to observation and inference is, really, far greater than that which attends any kind of skill or dexterity, even when death and pain add their zest to the latter. Let anyone who has an eye and a brain (but especially the latter), lay down the gun and take up the glasses [opera glasses, proto-binoculars] for a week, a day, even for an hour, if he is lucky, and he will never wish to change back again. He will soon come to regard the killing of birds as not only brutal, but dreadfully silly, and his gun and cartridges, once so dear, will be to him, hereafter, as the toys of childhood are to the grown man.[2]

Hated and derided by the museum experts – who considered themselves the *real* scientific ornithologists – Selous made empathy for birds respectable and, in doing so, changed the world. Birdwatching became one of the most popular pastimes worldwide, eventually making birding scientific and playing a pivotal role in their conservation.

After tossing aside his gun, Selous's approach was to watch, as he had done with the nightjars, and to *think* – think about why birds were behaving in particular ways. *Inference*, he called it. If his writing sometimes seems a bit convoluted it is partly because he was breaking new ground. There was no template, no mentors and no model on which he could base his ideas.[3]

Born into a wealthy and talented family – his father was chairman of the London Stock Exchange – of part-Huguenot heritage (the name was originally Slous), Edmund Selous was privately educated and spent a year at Pembroke College in Cambridge before training as a barrister. He gave up soon after qualifying to concentrate on natural history and writing. His older brother, Frederick, was a larger-than-life character who travelled to Africa aged just nineteen and became famous as a big-game hunter, killing elephants and providing specimens for museums. I have often wondered whether Edmund's rejection of killing-for-science was a reaction to his brother's brash, ego-boosting behaviour. This fits with the view that birth-order shapes personality as offspring compete for the equivalent of unique ecological

Edmund Selous, pioneering ornithologist of the early 1900s (courtesy K. E. L. Simmons).

niches. Frederick, the oldest son assumed the prime hunter-killer niche, while Edmund, to be distinct, took the gentler, more reserved birdwatcher niche.[4]

A Darwin enthusiast, Edmund Selous's discerning observations of Ruffs at their display grounds in 1906 provided startling evidence for an idea many of Darwin's critics thought implausible – that females choose their mate, rather than the other way round. Enthusiastically writing up his findings, Selous soon attracted the attention of others interested in birds. One of these was Henry Eliot Howard, sixteen years his junior, and a businessman who got up before dawn to watch warblers before he went to work. It was from these intimate and careful Selous-inspired studies that Howard came to recognize the central importance of territory in bird life, describing his discoveries in 1920 in a book that was later acknowledged as one of the most important biological works of the twentieth century.[5]

Also influenced by Selous and Howard was Julian Huxley, grandson of Darwin's bulldog, Thomas Henry Huxley. Like his two predecessors, Huxley observed birds closely in an effort to interpret their behaviour. His most famous study,

Ruff (female left, male right), a polygynous and strikingly sexually dimorphic species (from Yarrell, 1843).

on the courtship habits of Great Crested Grebes, was conducted in 1912 during his honeymoon – to the dismay of his bride. While probably not doing a great deal for their relationship, this study reflected his passion for birds as well as providing an important foundation for the study of animal behaviour.

It is difficult to overestimate the seismic shift in our relationships with birds that Selous, Howard and Huxley precipitated. Dispassionate killing was gradually displaced in favour of a gentler, more intimate approach whose aim was to better understand the nature of a bird's world. It was a shift enhanced, of course, by the appearance of decent binoculars, which in the early 1900s enabled watchers to observe birds from a distance without disturbing them.[6]

Another bird killer and collector who embraced the changing attitude towards birds was Harry Witherby. From a family whose business was publishing, Witherby saw the increasing interest in birds as a commercial opportunity. In 1907 he launched the monthly magazine *British Birds*, catering specifically to those whose ornithological horizons extended beyond collecting eggs and study skins. In contrast, *The Ibis*, the British Ornithologists' Union's much older and staunchly traditional journal, continued to publish articles mainly by museum ornithologists or those in different parts of the world obsessed with lists of species seen – and shot. Bizarrely, despite its claim to further the cause of bird conservation and biology, *The Ibis* remained predominantly museum- and list-based, right up until the 1940s – thanks to its deeply conservative editors.[7]

The observation of live birds was becoming increasingly popular, and Witherby's *British Birds* was a winner. Keen to capture more subscribers, Witherby boosted the magazine's appeal using photographs of birds, including some taken by the talented and pioneering Emma Turner. The magazine covered a wide range of bird-related topics but with particular emphasis on rarities, now more often watched than shot. Witherby also solicited submissions from eminent ornithologists including Horace Alexander, whom he asked in 1914 to write an article on the 'ecology' of birds, at that time a startlingly original theme. Alexander obliged, and his paper elicited an enthusiastic response from Witherby, who wrote to him suggesting ways the subject might be developed. But Alexander demurred, 'dried up' he said, by the horrors of the First World War. Undeterred, Witherby then wondered whether Eliot Howard might be the person to do this, but dismissed the idea almost as quickly because Howard, he felt, was so 'infernally theoretical'. Ecology did not die, it simply took a few more years to come of age.[8]

Harry Witherby was a mover, a shaker and a major player in the development of birdwatching. He masterminded the *Handbook of British Birds*, published between 1938 and 1941. Before that, however, he started one of the first bird ringing (banding) schemes, giving new impetus to birdwatching and our understanding of that mystery of mysteries, bird migration. Employing many of the techniques first used by the ancient Egyptians, bird ringers captured birds in order to place uniquely numbered metal rings on their legs. Trapping was both utilitarian and emotive, tapping into both a deep-seated hunting instinct and the satisfaction of success, but at the same time providing intimate contact with birds. A living bird in the hand is worth more than one viewed through binoculars, or dead: you can feel its heart beating, sense the texture of its plumage, appreciate the intricate arrangement of the feathers on its wings and tail and you can look into its eye. In most cases it is wonderful, but rather less so if it is a woodpecker that punctures both your daydreams and your hand in its bid to escape. When my children were small I caught a female Eurasian Sparrowhawk in the garden and, holding it to avoid its talons, I took it to show them. The bird's anger was all too evident from its piercing yellow eyes and sharp-clawed twitching talons. My kids were horrified – I think it put them off birds for ever.

Over successive decades of the twentieth century, ringing and migration became major aspects of bird study, with

Sparrowhawk – 'the hedgerow thief' – with its sparrow prey, painted by Jemima Blackburn in the mid-1880s (courtesy Alan Blackburn).

birders identifying offshore islands like Fair Isle and Skokholm as especially good places to see migration in action. These were also places where you could find and observe rare birds. In 1927 Ronald Lockley and his wife Doris moved to the uninhabited island of Skokholm in the southern Irish Sea to live among its breeding seabirds and seasonal surges of migrants, there to start Britain's first bird observatory.[9]

Through a swift succession of books applauding the works of Selous, Howard and Huxley in the 1920s and 1930s, Max Nicholson emerged as Britain's foremost birdman. Despite his predecessors' curiosity-driven birding, Nicholson himself was committed to the idea that birding should be useful, and went on to launch several groundbreaking national surveys of bird abundance.

A second boost to birdwatching came in 1940, in the early days of the Second World War, with James Fisher's *Watching Birds*, a book that eventually sold over a million copies, and which, in its introduction, emphasized the variety of people then engaged in the hobby. For those directly involved in the war, watching birds was a welcome distraction during the long, boring intervals between fighting. And for those confined in German prisoner-of-war camps, birds provided a much-needed antidote to boredom and despair. One of the servicemen who took solace in birds was John Buxton, who had served as warden of Skokholm Bird Observatory in 1939 with his wife Marjorie, Ronald Lockley's sister.

Captured in Norway in May 1940, Buxton spent the rest of the war in various prison camps, where he encouraged fellow inmates to watch and record the behaviour of the different

John Buxton, head of table, with his wife Marjorie to his right in the Wheelhouse, Skokholm Bird Observatory, August 1939 just weeks before the start of the Second World War (courtesy S. Sutcliffe).

birds they could see. From his camp, Buxton wrote to Erwin Stresemann, Germany's leading ornithologist, who in a wonderful gesture of collegiality responded by sending books and bird rings to help with their studies. A true scholar – Buxton had been partway through an Oxford DPhil at the outbreak of war – he subsequently transformed the mass of notes accumulated by his fellow prisoners into a wonderful monograph on the Common Redstart, published in 1950.[10]

A male Common Redstart, the species that helped keep British POWs occupied during the Second World War (from Yarrell, 1843).

Once the war was over, interest in birds metamorphosed from a 'comparatively rare eccentricity into a national pastime', and as it did so Max Nicholson's carefully woven tapestry of 'useful ornithology' started to unravel.[11] Two increasingly distinct strands – that he would have identified as either purposeful (censusing) or aimless (literally – birding) – were starting to emerge: surveys versus listing. In a way, such a divide was inevitable. As more and more people became interested in birds, it was unrealistic to expect them all to engage in something 'useful'.

The tension created by this developing division was aired in some bird journals. Reading *British Birds* as a PhD student on Skomer in the 1970s, I was puzzled by the fact that early in its life it had published papers written by mainstream scientific ornithologists, whereas from the 1960s such papers were rare and the emphasis was increasingly on identification and the occurrence of rarities. This emerging ornithological watershed generated some strong feelings, exemplified by the Reverend Peter Hartley, a professional ornithologist, who in 1954 declared that 'Non-scientific birdwatching . . . is simply lazy, incompetent and slovenly birdwatching.'[12]

In response, Denis Summers-Smith, an amateur who eventually became Britain's grand old man of sparrow research, countered by saying that birdwatching 'is no more slovenly . . . than going to a concert without a score . . . many are not suited to carry out scientific studies or read scores. Should we criticize them for the pleasure they get from birds or music?' He finishes by saying: 'Let's have no more of this scientific snobbishness: a real enthusiasm will have more influence in attracting others . . . than any amount of castigation.' Hurrah![13]

Hartley's histrionics were actually a clumsy attempt to encourage the amateur or 'back-garden ornithologist', who, because he 'sees the same birds every day and for many days is able to make those systematic observations which are the very basis of scientific ornithology'. Digging himself further into a pit, Hartley tried to reassure his readers that 'the characteristic of the scientist is neither cold detachment nor a love of the abstruse, but rather an enthusiasm so furious that he can regard no detail as unimportant and no exertion in the field or in the study as unwarrantable'.[14]

Known for his prickly nature, Hartley was hardly a role model, nor can I imagine his depiction of the professional ornithologist encouraging amateurs to set to work. A contemporary, Edward Armstrong, another churchman, more benign than the belligerent Hartley, did a better job of convincing amateurs that they could contribute to ornithology. He suggested that they might provide detailed descriptions of bird displays and offered a list of species for which such information was missing. An amateur, and author of *Bird Display and Behaviour* (1942), Armstrong was inspired by the rapidly developing field of animal behaviour and by the fact that its star academic players, Konrad Lorenz and Niko

Tinbergen (whom we will meet in the next chapter), had also produced popular accounts of their research.[15]

Nicholson's strand of 'purposeful birding' led to the formation of the British Trust for Ornithology (BTO) in 1932, whose primary role was the assessment of bird numbers and whose survey results initially found a home in Witherby's magazine. Such was the success of this approach that, by the 1950s, it was suggested that *British Birds* might become the official organ of Nicholson's British Trust for Ornithology. But the increasing division between purposeful and less-than-purposeful birding reminded the editors of *British Birds* that not all its readers relished the often tedious (but essential) details of the trust's scientific surveys. It was felt better that they should start their own journal and the result was *Bird Study*, first appearing in 1954, a publication that allowed *British Birds* to realign itself as the birders' journal focusing on bird identification and rarities.[16]

The battle between Hartley and Summers-Smith was symptomatic of birding's ongoing bumpy descent through the social hierarchy. In the 1800s and early 1900s only the wealthy could afford a serious interest in birds. Even by the 1950s birdwatching continued to be dominated by those 'that held sway in most departments of cultural life' in Britain – that is, mainly upper-class white males. As interest in birds continued to expand, by the 1970s and 80s most birders 'came from the same broad social background – the working and middle classes'.[17]

There have been several attempts to categorize the different types of people who engage with birds. The writer Mark

Cocker, for example, likened bird lovers to a tribe among which an amateur anthropologist such as himself could identify eight separate sub-clans: (1) scientist, (2) ornithologist, (3) bird-watcher, (4) birdwatcher, (5) birder, (6) twitcher, (7) dude and (8) robin stroker. Of these, he says, scientist and ornithologist overlap almost completely; the next four categories have much in common too, although Cocker is at pains to distinguish the bird-watcher – which he identifies as being like the train-spotter – from the birdwatcher, preferring unhyphenated authenticity, or better still, birder, the term now in most common use. A twitcher, he says, is someone who, as we will see, pursues rarities, and is often considered 'the most evil creature alive . . . fanatical, self-indulgent, uncaring, competitive and anti-environmental'. Dudes are posers, pretending to know more about birds than they do and universally despised by those above them in the avian hierarchy. The final sub-clan is what Cocker calls the 'robin strokers' – those people who watch birds from their living-room window. These, he says – in case anyone thinks his term disparaging – comprise 'the vast bulk of decent folk without whom bird conservation would have no real teeth'.[18]

As a scientist, I'd prefer a broader, more objective assessment that included social class, gender, ethnicity and age. Cocker's eight sub-clans might be a start – although would anyone ever admit to being a dude or a robin stroker?

A major difficulty in classifying people's interest in birds is that none of the categories is exclusive. It would be so much simpler if they were, but here I am, a scientist, thrilled by the (wild) European Robin that has just flown in – as it often does – through the open door and perched above me as I type.

So what can we say? In nineteenth- and early-twentieth-century Europe, North America and Australia, a serious interest in birds was predominantly a white, wealthy, middle- or upper-class and often-colonial male preserve. It was, as we have seen, largely based around the acquisition of eggs and bird skins. Less well known is that at that time almost every second household across Britain and the Continent held one or more cage birds cared for by both men and women. The start of birdwatching in the early 1900s, while retaining its scientific remit, fostered a change from killing to caring, but – with some notable exceptions – was still pursued principally by men. Over succeeding decades birding became more genteel: my early days of skulking in the undergrowth with my binoculars are a thing of the past, for most birders today are conveyed along wooden boardwalks towards cosy hides to watch birds in comfort. Since the 1960s, the ongoing expansion of higher education across much of the world has resulted in more and more women taking university courses in biology and zoology and becoming professional ornithologists – although senior positions are, sadly, still mainly occupied by men. Among birdwatchers the proportion of women has also increased, and I was told in 2020 that in Australia women birders now outnumber men. The next challenge, of course, is to provide more opportunities for ethnic minorities to develop an interest in birds.[19]

When I was very young my father took me birdwatching, allowing me to look – always put the strap round your neck! – through his Zeiss Dekarem 10×50 binoculars. When I was five, and my brother Mike three, he gave us each one of Eric Pochin's books: mine was *How to Recognize British Wild Birds* and my brother's was *More About British Wild Birds*. Their unusual and simplistic images still trigger wonderful childhood birding memories. I'd like to think that my dad had read Nicholson's books, but I doubt it, and I regret now never asking how he became interested in birds. The interest that he had fostered soon became an obsession, and on some days, instead of going to school, I went birdwatching. Apart from biology and art, there was nothing that interested me at my school, least of all its almost total preoccupation with sport and military games. I was considered a no-hoper: useless on the sports field, scathing of the banality of playing soldiers and an almost complete academic dead loss whose rebellious behaviour elicited increasingly severe punitive measures. Then one day my dad took me to one side and said, 'You know, if you don't pass these exams [O levels] you'll end up working in a cloth factory in Leeds.' It did the trick and I began to focus on my school work. I continued to birdwatch, but only at appropriate times. Rather like the epiphany experienced by Selous, I underwent one of my own and discovered that I liked learning stuff.

Goodness knows where that came from, as neither of my parents was the slightest bit academic, and both had left school at some ridiculously early age – my dad, growing up in Hull, ran away to sea and worked on trawlers in the Barents Sea, and my mum became a secretary at the Colman's mustard factory in Norwich. My dad was a classic type-A personality – goal-orientated and an achiever – and my mum, one of life's enthusiasts, was an accomplished amateur artist. Those were the traits, via nurture and nature, that prodded me down the path towards a career in ornithology rather than something else in which birdwatching would have been merely a pastime. As it was for many of my ornithological colleagues, birdwatching was the route to my scientific career, and has remained a hobby alongside it.

Reflecting on my own experience, I started to think about what it is that encourages a young person interested in birds towards or away from becoming an academic ornithologist – someone who studies birds scientifically. After talking to various colleagues, I came to the conclusion that it boils down to three things: opportunity, luck and curiosity. In my own case the opportunity was having been sent to a 'good school', which, although I hated it, did instil in me a kind of ambition. My luck was the mid-1960s change in the university system that allowed someone with no language qualification and only two sciences (and – how bizarre! – art as an A level), to undertake a degree in zoology. A second piece of luck was having lectures from Robin Baker – someone who understood natural selection and told us about its looming possibilities in reshaping biology. As for curiosity, from an early age I wanted to know more about birds than listing them. I have no idea where that came from. The Nobel Laureate Peter Medawar, a biologist famous for his work on immune tolerance and tissue transplants, similarly identified curiosity as a key trait among biologists. He said – correctly in my case – that one did not have to be especially clever to be a scientist, but 'common sense one cannot do without'. One also needs 'diligence, a sense of purpose, the power to concentrate and not to be cast down by adversity'.[20]

There's another aspect of curiosity I find intriguing. Some of my scientist colleagues talk about being *hungry* for information and data (as well as research funding and presumably success), so perhaps it is not too surprising that those regions of the brain responsible for our desire for food are the same as those controlling our desire for knowledge, since both activities are forms of reward-seeking.[21]

On the other side of the coin, I have also wondered what deters or prevents a teenager interested in birds from becoming an ornithologist. The ongoing lack of ethnic minorities in science and ornithology (to say nothing of birding itself) clearly demonstrates the effect of opportunity. Another aspect of opportunity is that prior to the 1970s, biology was considered science's softest subject, and if one was both smart and scientifically inclined, mathematics, physics and chemistry (often leading to a career in medicine) were the way forward. Certainly, some of the birders I have spoken to told me that they were discouraged from a career in ornithology because its prospects seemed so poor. That changed after about 1970, coinciding with the coming of age of behavioural ecology, and most university zoology departments wanted a behavioural ecologist on their staff.

Of the many influences that shaped my life, *British Birds* magazine played a major role. During my time as a research student on Skomer Island in the 1970s the only reading matter was a complete set of *British Birds* – seventy volumes – that someone had donated to the island. The early part of the guillemot breeding season is characterized by the birds' alternating presence and absence at the colony, the absences often coinciding with bad weather, rendering such days rather tedious. To stave off boredom, I read all the issues of *British Birds* and could probably have entered *Mastermind* with that as my specialist subject. It was the 'Notes' – brief accounts at the back of each issue – that were the most intriguing. These were often sightings (occasionally, shootings) of rare birds or unseasonal migrants (such as a Brambling in June, a Fieldfare in August or a Barn Swallow in December), but also unusual observations such as a case of several geese and gulls being killed by lightning, a Great Auk bone found in the Broch of Mousa, Shetland, or four adult Long-tailed Tits feeding chicks in the same nest. Reading those copies of *British Birds* gave me a strong sense of the wonderful diversity among both birds and birders, and those quirky Notes made me realize how that scattering of anecdotal observations sometimes contained the germs of new developments in bird studies.[22]

Of all the different styles of birding, extreme listing – twitching – is the most remote from my own interests. I like seeing new birds, but I have rarely travelled any distance to tick a rarity, although I have ticked quite a few twitchers.

Sitting on the front row of a set of seats in a church hall, I am psyching myself up to give a talk. As people file past and take their places, my neighbour, a middle-aged woman I do not know, nudges me in the ribs and with extraordinary vehemence spits out the words: 'That's disgusting!' I have no idea what she is referring to, but seeing my bewilderment she nods aggressively towards a man and a woman taking their seats behind us. The man is someone I recognize – Jon Hornbuckle, a local birding legend. The woman is his new partner. I presume it is their age difference that has so incensed my neighbour. I felt honoured that Jon, whom I had first met many years previously, should be sufficiently interested to come to my talk.

Having seen more species of birds across the world than anyone else, Jon was Britain's top lister. As such, he had reached the summit of the listing movement that began in the United States in 1900 when Frank Chapman launched the Christmas Bird Count. Chapman's Bird Count, in which participants recorded all the bird species they saw, was an alternative to the Christmas Side Hunt, a 'festive slaughter' in which participants competed to see how many animals they could kill on Christmas Day. Increasingly popular, the Christmas Bird Count eventually eclipsed the Side Hunt to become an early citizen science project that, as well as providing lists of all birds seen by birders across the US, was the beginning of bird monitoring and protection that continues to this day.[23]

Listing is at the heart of birding, and all the birders that I knew as a boy listed what they had seen on each outing. The list is a personal record and an opportunity to compare notes with others, sometimes providing some friendly – occasionally not-so-friendly – competition. In a few instances listing mutated into the meme-equivalent of an auto-immune disease, eating its owner from the inside out, as it was eventually to do for Jon Hornbuckle.

When I first arrived in Sheffield to take up a position of lecturer in zoology at the university in 1976, the Sheffield Bird Study Group was just four years old and exuded extraordinary youthful vigour. At that time Jon, a senior metallurgist with British Steel, was a driving force in the bird group, serving successively as secretary, chairman and editor of an annual report that in 1991 won a national award. Exceptionally dynamic, Sheffield's Bird Study Group encouraged young birders, conducted surveys – of Eurasian Magpies, Rooks and the Common Starlings that then roosted in vast numbers in the city centre – and protected raptors from rapacious gamekeepers.

Some of the group's younger members were undergraduate students of mine, one of whom came to tell me with great excitement that he had just discovered a pair of breeding Peregrines in the Peak District not far from Sheffield. In the 1970s this was remarkable, for the species was still extremely rare after its pesticide-induced decimation. The student told me exactly where the nest was so I could go and see the birds for myself. No sooner had he left my office when another young birder came in to announce the same news. Feigning ignorance, I asked where the nest was. 'I cannot possibly tell *you*,' he said, but instead showed me a 35mm transparency of the nest site as evidence of the find. Continuing the deception, I told him that after years looking at guillemot cliffs I had a photographic memory for rock faces and could probably figure out from the geology on the slide where the Peregrines were nesting. He cast me a look of disbelief. And then, just as in Roald Dahl's short story 'Taste' – in which a dishonest dinner guest bets his host that he can identify the vineyard from which a particularly rare wine originated – I slowly homed in on the exact location of the Peregrine's nest, to the student's dismay and incredulity. It was only years later that I told him the truth.[24] Hornbuckle and his colleagues mounted twenty-four-hour watches on that nest to ensure that the chicks fledged successfully. Those birds were the front-runners of an expanding Peregrine population, some of which now nest in Sheffield's city centre and are visible from my office window.

Hornbuckle's obsession emerged from a boyhood hobby of train-spotting. His conversion came during a family holiday on the Yorkshire coast in 1970, when he visited Bempton Cliffs and was fascinated by the newly formed Northern Gannet colony, which then comprised just six pairs (there are now around 11,000). The next year a steel industry award sent Jon to Japan on what became a

Peregrine Falcons: young (upper) and adult (lower) from Harting (1901).

three-month round-the-world birding tour, setting the course for the rest of his life. Taking early retirement in November 1993, he left the next day for Ecuador. Over the following twenty-five years Jon travelled the world, clocking up a birding list longer than that of anyone else, claiming to have seen 9,600 of the world's 10,000 or more known species. Only a handful of other birders have come close, including the better known Phoebe Snetsinger, who started birding in her fifties after being diagnosed with cancer, and who saw over 8,000 species before dying eighteen years later in a car accident while birding in Madagascar. As well as being the ultimate lister – a passion that cost him his first marriage – Jon was a keen conservationist and helped to protect birds in various parts of the world. He too died following a road accident, in 2018.

Jon's story spans a period of rapidly changing attitudes towards birds. From the 1960s international travel became cheaper and easier, eco-tourism became seductively popular and field guides arrived like flocks of winter waxwings, allowing birders to familiarize, memorize and fantasize about the birds they might see on forthcoming trips. The business of bird identification and verification was transformed by binoculars and telescopes of ever increasing clarity, and digital cameras and smart phones made everyone a photographer. Birders became increasingly sophisticated in their identification skills, even to the point of discriminating between different races of Fox Sparrows in North America and chiffchaffs in the UK. These are all changes reflected in Britain's annual Bird Fair, an event attracting thousands of UK birders of all types and providing a model that has since been copied across the world. Part of the Bird Fair's deep appeal is that its profits go to global bird conservation. The other aspect is that it reinforces the friendships and camaraderie that a shared interest in birds generates. Once the preserve of loners like Selous, and, indeed, myself, birding is now a much more social activity.

Obsessive listing or twitching differs only from the obsessive collecting of eggs or skins in its legality, and like these activities, hovers precariously at the far end of the spectrum of our bird-related interests.[25]

Worldwide, tens of millions of people have an interest in birds. Because there's no precise definition of what a birder is, there's no precise figure. It is telling, however, that the Royal Society for the Protection of Birds in the UK has more members than all UK political parties combined. Current interest in birds is fuelled by the recent spike in nature publishing and in wildlife television programming – most notably by Sir David Attenborough's status as a secular saint. This interest is a powerful economic force, creating a market for increasingly sophisticated (and expensive) equipment and clothing, bird books and bird food, as well as ecotourism. A recent North American survey estimated that birding is worth around $96 billion annually, supporting around 700,000 jobs and generating $17 billion in tax revenue.[26]

The irony of all this is that as the number of people interested in birds has boomed, the numbers of birds has bombed – more and more people are in search of fewer and fewer birds.

The good news is that the increase in the number of birders, together with some stunning technological developments like eBird and ICARUS, are transforming many different forms of birding, giving them purpose and enhancing our knowledge of bird biology on a scale no one could have imagined even twenty years ago.

Created by the Cornell Laboratory of Ornithology in 2002, eBird is an online global database that documents bird sightings, and so far half a million birders have logged almost one billion individual sightings. Records are submitted through an app that logs the user's location and permits rapid record sharing. It sounds almost too good to be true, but since 'rivalry, animosity and ego have long been hallmarks of the bird world . . . a code of ethics is necessary'. The announcement on eBird of a rarity in someone's garden or back yard, for example, can result in the appearance of large numbers of birders, not all of whom exhibit the expected etiquette. Another issue is identification. Quality control is a key question with all types of citizen science. Distinguishing a Dickcissel from a meadowlark can be tricky for a beginner, but eBird addresses identification issues by signing up as reviewers experienced birders, who scrutinize submissions and tactfully rectify errors.[27]

eBird is an incredibly powerful tool that brings birders and scientists closer together than ever before. Through the careful analysis of eBird records it is now possible to track the global migration routes of entire species with unprecedented accuracy; it is also possible to tell whether populations are increasing or decreasing, and it sometimes even allows us to witness hitherto unknown aspects of their biology – as we will see in the next chapter.

11. A Boom in Bird Studies: Behaviour, Evolution and Ecology

> It was at Clark University that I found a purpose in life . . . at every turn there was a challenge – nature waiting to be studied and understood. For the first time I comprehended the phrase . . . 'the search for truth' . . . to follow hidden paths and find unsuspected beauty and truth . . . to venture hypotheses as to the why.
>
> Margaret Morse Nice (1979)

My passion for birds as a child led me to hand-raise several abandoned young Eurasian Magpies, Western Jackdaws, Rooks, Tawny Owls and Common Starlings. Anyone who has done this will know what hard work it is, since, like human babies, chicks require feeding every few hours. Yet rearing *thousands* of young birds, often from the egg, is what Magdalena Heinroth did over a period of almost thirty years in the apartment she shared with her husband Oskar, who was Director of the Berlin Zoo's aquarium.

Born in 1871 into a family of musicians and academics, Oskar Heinroth was precociously clever. At the age of four he could recognize all the chickens in the family hen house – simply from their voices. Growing up, he built an aviary in his bedroom to house native birds. As was common at that time, he trained as a medic, but in 1896 went to Berlin to study zoology. An invitation by a wealthy private sponsor took him on a collecting expedition to the Bismarck Archipelago off New Guinea in 1900. When visiting the St Matthias Islands, the pair were attacked by islanders and Heinroth's sponsor was killed. Heinroth, who had been lying in his tent sick with malaria, escaped with a spear wound to his leg, an injury that left him with a permanent limp. On his return to Berlin he wrote up and published his findings from the expedition, providing the first inklings of a novel approach to the study of birds. Just as people had previously assumed Aristotle to have discovered everything worth knowing about natural history, ornithologists in central Europe assumed that Johann Andreas Naumann and his son Johann Friedrich had done the same for birds in the late 1700s and 1800s, and that there was really nothing new to be learned. Heinroth rejected this, and was duly credited with 'breaching the walls of the Naumann cult'. Even though his first major foray into ornithology had been a classical collecting trip, he soon adopted an entirely new, empathetic approach to studying birds.[1]

Berlin Zoo's collection of waterfowl – ducks, geese and swans – was Oskar's passion. For several years he watched and recorded their courtship behaviour with a critical and judicious eye. Showing great insight, he realized that the displays of different species seemed to be hard-wired, or innate, and could, like the anatomical features traditionally used by ornithologists, shed light on the birds' evolutionary history. This finding formed the beginnings of a bridge between avian systematics (how birds evolved) and the natural history of birds. But it did more than that, for Heinroth's study of ducks proved to be one of the most important foundations in the developing study of animal behaviour in the natural environment.[2]

When Oskar proposed to Magdalena in 1902 he presented

her with a Eurasian Blackcap as a gift. They married two years later, but she was unable to conceive, probably as a result of gynaecological surgery she had undergone in her twenties. Dissatisfied with the domesticity of married life, Magdalena started to care for birds instead of babies. She and Oskar bought two young European Nightjars – in very poor condition – from a bird market and nursed them back to health. Was their choice of the nightjar, I wonder, inspired by Edmund Selous's compelling 1899 account of this species in *The Zoologist*, just a few years earlier?

The first decade of the 1900s saw interest in the behaviour and ecology of birds start to emerge like random fiery patches of rosebay willowherb in a bombed-out landscape of killing and listing. Those fiery patches began to appear all over Europe and North America, prompting Heinroth to ask how much of a bird's behaviour was due to nature and how much to nurture. With Magdalena's wonderful facility for rearing birds, he saw a new horizon opening up.

Addressing this most fundamental of biological questions led to a twenty-eight-year-long project in which Magdalena raised dozens of different bird species – a project that started as a hobby but became an obsession. Few individuals have had such close, intense and intimate relationships with birds. Helping them hatch was like giving birth and, like a new mum, Magdalena cared for and fed her babies day and night, watching anxiously to ensure that they were putting on weight and then rejoicing in their first steps or wing shakes. Through beautifully composed photographs, Oskar and Magdalena documented the way different species develop from chick to adult, assessing how nurture and nature seemed to mould each bird's behaviour. Almost three decades and 286 species later – ranging from tiny Goldcrests, to gigantic

Lammergeiers – Oskar and Magdalena had published four fat volumes describing their findings.[3]

No one had done anything like this before. Readers were impressed by the scale of the achievement, and especially by Oskar's photographs, though rather less so by his hastily written text, composed at the kitchen table in the evenings after long days at the zoo. Sadly, the results of this study, one of the most remarkable of twentieth-century biology, failed to find their way into mainstream ornithology. It was a matter of timing and language. Written in German, Heinroth's text was inaccessible to most English-speaking ornithologists and, despite his attempt to unravel the Gordian nature–nurture knot, Heinroth failed to draw any *general* conclusions from the study. Instead, his inferences were hidden in the individual species accounts, making it difficult for readers to see any

Oskar and Magdalena Heinroth in 1904 – the year they married (courtesy of Klaus Nigge and Staatsbibliothek zu Berlin).

overall patterns. Little wonder, then, that Heinroth's four volumes, notwithstanding their bulk, slipped irretrievably between the cracks.

One year, Magdalena hatched a clutch of twelve Corncrake eggs in an incubator, naming the last emerging chick Lilliput on account of his small size. Remaining as a family pet for several years in their apartment, Lilliput drove the Heinroths' neighbours to distraction with his ceaseless and far-carrying *crex-crex* each breeding season. Lilliput was in love with – imprinted on – Magdalena, revealing his passion by attempting to copulate with her hand.

Another year it was Great Cormorants and Eurasian Spoonbills, acquired from the Netherlands. Knowing that the parent birds fed their chicks on regurgitated fish, this is what Magdalena did too: masticating the raw fish into a mush, and transferring it directly from her mouth to theirs, much as the indigenous South American women we saw in Chapter 6 did when hand-rearing parrots, but presumably, for her, less pleasantly.

To obtain Eider eggs, Oskar and Magdalena travelled to Sweden, returning to Berlin on the overnight train (first class), keeping all but one of the eggs cosy in a foot warmer during the journey, the other being 'incubated' in Magdalena's cleavage. This was the egg that produced Edda, assumed initially to be female but who proved on maturity to be a male and who subsequently became the forefather of Berlin Zoo's entire colony of Eider Ducks.

By 1932, with the project complete, Magdalena took a much-needed holiday with friends on the Danube, leaving Oskar in Berlin. The previous year she had been diagnosed with breast cancer, but had recovered well from a double mastectomy. On holiday, however, there was a complication, arising not from the cancer but from surgery she had undergone as a young woman. Her condition was serious and required immediate attention. Alerted by telegram, Oskar set off for Budapest by express train, but too late: Magdalena died of a perforated bowel on 15 August. With the coffin strapped to the roof of a car, Oskar took his beloved wife's body to Budapest's crematorium and then returned to Berlin by train, the urn containing her ashes in his lap.[4]

Oskar remarried within a year. His new wife, the very capable Katharina, who had previously worked with the scientist Karl von Frisch on honeybees, was, like Oskar, a bird lover. The 1930s proved to be a productive and happy time for Oskar and Katharina. Despite the deteriorating political situation in Europe, Oskar continued to organize weekly bird seminars at the zoo with his good friend Erwin Stresemann.

Stresemann was Germany's leading ornithologist. In 1914, aged just twenty-five, he had been invited to write a handbook of birds for a series of zoology texts, and then by 1921 he was editor of Germany's premier bird journal *Journal für Ornithologie*. By encouraging papers across a broad range of topics, including field-based studies of the ecology and behaviour of birds – subjects still quite alien at the time to the majority of ornithologists – Stresemann shaped the way ornithology developed in central Europe, making it part of mainstream biology.

A single example illustrates Stresemann's far-sightedness. During the 1920s Margaret Morse Nice conducted a detailed study of the behaviour and life history of Song Sparrows near her home in Ohio but was unable to get her work acknowledged, let alone published, in her own country. Recognizing the novelty and quality of her work, Stresemann arranged for her to travel to Germany in 1932 and for her studies to be published in his journal.[5]

Stresemann was said to have 'opened the windows and doors, permitting fresh air to blow through the halls of ornithology'. Delayed by the First World War, his four-volume *Aves* finally appeared between 1927 and 1934. It was an absolute tour de force and, had it been translated into English, it would have been *the* landmark bird book of the early twentieth century.[6]

Through the 1930s, Heinroth and Stresemann must have despaired as they watched the situation in Germany descend into war. Berlin became the inevitable target of Allied bombs, and by 1945, with much of the city's infrastructure gone and most of the zoo's animals dead, Oskar lay, undernourished and riddled with infection, in a damp cellar. He died three weeks after the war ended.[7]

Although now largely forgotten, the Heinroths' ornithological efforts, like those of Selous and Howard in England, had the remarkable effect of converting collectors into biologists. Deservedly, the Heinroths have enjoyed a recent renaissance. Karl Schulze-Hagen, a full-time medic and part-time ornithologist based in Mönchengladbach, has worked hard to publicize their achievements. Inspired by the images in the Heinroth's book, Karl visited the Berlin Staatsbibliothek in the 1980s, wondering whether there was a Heinroth archive. There he found thirty-five boxes containing letters, diaries and hundreds of tiny contact prints of Heinroth's wonderful photographs, all untouched since the 1940s. Hoping that at least a few of the original photographic plates might have survived the war, Karl continued his search, this time in the archives of Berlin Zoo. It took three years of polite and persuasive persistence before, in 2019, he was finally allowed into the zoo's cellars, where he found 800 glass plates, long since thought to have been destroyed. These images included the Heinroths' hand-reared goshawk that killed and ate a (hand-reared) pheasant on their balcony, the nightjars that nested on their dining-room floor and the two cranes they walked in the zoo's grounds. Here was the path to the Heinroths' resurrection, those pictures providing some of the best evidence of an extraordinary, scientifically motivated empathy for birds that captured the spirit of the twentieth century's new ornithology.[8]

The studies undertaken by Selous, Howard, Huxley and the Heinroths provided the impetus for a move away from museum ornithology to the investigation of living birds. They were the foundation of a new field of animal behaviour whose main architects were Konrad Lorenz, Niko Tinbergen and Karl von Frisch, with Lorenz and Tinbergen focusing mainly on birds.

Good friends and both brilliant in their own ways, Tinbergen and Lorenz had the common goal of establishing a new discipline. But as people they could hardly have been more different. Lorenz was (and remains) controversial. A keeper of animals from a young age, he quickly saw the potential in Heinroth's approach when he received the first of his four volumes in 1925 as a gift, and subsequently stood – rather heavily, it must be said – on Heinroth's shoulders. It is often difficult in science to distinguish plagiarism from inspiration, but in this case Lorenz seems to have both assumed ownership of and developed many of Heinroth's ideas. He was hugely ambitious, Heinroth less so. Lorenz acknowledged him, but being an enthusiastic self-publicist, it was Lorenz who subsequently received the accolades. Keeping animals allowed the Heinroths and Lorenz to study their behaviour at close range, providing the opportunity to interpret what they saw, much as any dog or cat owner does. But it was Oskar's earlier discovery that the innate courtship behaviour of waterfowl could reveal their evolutionary history that really piqued Lorenz's interest. Another source of inspiration for Lorenz came from the Estonian biologist Jakob von Uexküll, who in the early 1900s came up with the idea that animals lived in a subjective world defined by their sensory systems, wherein only certain things matter: the *Umwelt*, as he called it. A swift spending much of its life in the air relies primarily on its eyesight and has a very different *Umwelt* from that of a nocturnal kiwi that depends more on touch and smell. It was precisely (or perhaps imprecisely) this understanding of a bird's world that Eliot Howard had unsuccessfully attempted to study. Lorenz picked up the idea, and in particular the notion that a crucial part of a bird's world was its interactions with members of its own species: its parents, its siblings, its partners and so on – a point forcibly brought home to him when his hand-reared and imprinted Western Jackdaw, Jock, began to courtship feed him as though he was its partner – rather like Magdalena's amorous Corncrake.

Tinbergen's approach was very different, studying wild birds (and other animals) as they went about their day-to-day business. Lorenz and Tinbergen first met in 1936, and the following year, Tinbergen and his wife Elizabeth went to stay with Lorenz and his family, forging a lifelong friendship. Their different approaches to the study of behaviour – Tinbergen the hunter; Lorenz the farmer – went hand-in-hand and the two men fed voraciously off each other's ideas as they started to construct the new discipline of animal behaviour – known then as 'ethology'.[9]

Ultimately, Lorenz and Tinbergen were intrigued to know whether the behaviour of birds like jackdaws or Herring Gulls could reveal anything about ourselves. The many similarities between birds and humans were certainly suggestive, and since you can hardly study a species' behaviour without growing to like them, this cemented the feelings they had for birds. The concept of *Umwelt* and an appreciation of what it was like to be a bird were central to their entire endeavour, and I have often wondered whether they applied that concept to themselves as they lived out the horrors of the Second World War on opposite sides. For two long years Tinbergen was incarcerated, torn away from his wife and young children by the Nazis after they invaded the Netherlands. Lorenz, in Austria, opportunistically joined the Nazi Party in the hope of landing a job under the new regime. Captured by the Russians in 1944, he was a prisoner-of-war until 1948. Despite their differences, the two men resumed their friendship on being reunited in 1949 and continued together to promote and develop the study of animal behaviour.[10]

A paper that Tinbergen published in 1963 brought everything together. His 'On Aims and Methods in Ethology' identified the four different ways one can study and understand an animal's behaviour; four different ways behaviour could – or should – be investigated: (i) how behaviour has evolved over time, (ii) its current survival and reproductive value (adaptive significance), (iii) how behaviour develops as the animal grows up, including whether it is innate or learned, and (iv) the underlying physiological causes of different behaviours. Together with Karl von Frisch, who worked out the dance language of honeybees, Tinbergen and Lorenz jointly received the Nobel Prize in Physiology and Medicine in 1973 for creating the study of animal behaviour.[11]

Nobel Prizes are awarded for research and discoveries that are often long past, and indeed, by 1973 Tinbergen and Lorenz's work was rapidly becoming outdated. I discovered this the hard way, for in my PhD studies I attempted – in classic Tinbergen style – to catalogue (and photograph) the guillemot's entire repertoire of social signals to give me a sense of its social *Umwelt*. It took a huge amount of effort and I was deflated when my PhD examiners dismissed that chapter of my thesis as 'old hat'. In fact, as I later discovered, the descriptions of displays performed by birds, stimulated by Tinbergen and Lorenz's approach, are actually timeless since they form the bedrock of the species accounts in such handbooks as *Birds of the Western Palearctic* and *Birds of the World*.

Descriptions of bird displays also helped to pave the way for a new approach to thinking about the survival and reproductive value of behaviour – the second of Tinbergen's four questions. The assumption was that if a behaviour exists then it must somehow increase a bird's chances of surviving

Some of the courtship displays and postures used by Herring Gull (from Tinbergen, 1960).

and leaving descendants. The challenge was to discover *how* a particular behaviour was adaptive. Figuring out the value of a display like the Herring Gull's 'long call' is tricky. Other behaviours, like the mobbing of foxes and hedgehogs by gulls, proved much more tractable. Through a series of elegant field experiments – Tinbergen's forte – he and his students showed that mobbing was adaptive because it helped to drive predators away from the breeding colony.

One reason why Tinbergen succeeded in identifying the value of mobbing was because this behaviour is a joint effort by most members of the gull colony. What Tinbergen did not address explicitly is how mobbing benefits *individual* gulls, or whether it might pay some gulls to avoid being caught by a fox during mobbing, by simply hanging back and letting others take the risk. These would have been tricky questions to address in the 1960s, since most biologists then did not think about natural selection operating explicitly on individuals. As a result, most studies of animal behaviour at that time merely paid lip service to the question of survival value.[12]

There's a deep irony here, for the building in Oxford where Tinbergen had his office was also where the bird ecologist David Lack worked. Unlike Tinbergen, David Lack had a crystal clear understanding of how natural selection worked. Had he and Tinbergen joined forces, the combination of Lack's evolutionary framework and Tinbergen's experimental approach would have led them to do even greater things. As it was, that would come from others.[13]

Lack was one of just a handful of biologists who used natural selection as the guiding principle in their research. He sometimes had to confront the challenge posed by others who thought differently. The most notable of these was Vero Copner Wynne-Edwards, Professor of Natural History at the University of Aberdeen. Precocious teenage letter-writer that I was, I had written to both Lack and Wynne-Edwards as I struggled to understand what my herons were doing (Chapter 6). Both men sent helpful replies; Lack's was just a few lines, but Wynne-Edwards's was two closely typed sides in which he told me about his own boyhood birdwatching, which, remarkably, took place in some of the same areas as mine. Those letters, which I still have, were immensely encouraging.

Wynne-Edwards's view of natural selection was that it operated on species or groups, suggesting, for example, that animals kept their numbers in check by refraining from breeding when it looked as though food might become scarce. He and Lack clashed – albeit in gentlemanly manner – on this issue on numerous occasions. Seduced by the intuitive (and anthropomorphic) appeal of an idea he felt was on a par with Darwin's, Wynne-Edwards failed to acknowledge that it was illogical. Lack spelled it out for him: genes that cause an animal to avoid breeding simply do not make it into subsequent generations.[14]

Wynne-Edwards stuck to his guns, bolstering his ideas with his massive *Animal Dispersion in Relation to Social Behaviour* in 1962. But this was his undoing, for the book incensed others committed to the idea of individual selection. The American biologist George Williams had been busy writing his own book on evolution when Wynne-Edwards's appeared. Reading it made Williams start again, in a much more emphatic manner, and his *Adaptation and Natural Selection*, published in 1966, became the beacon for what was to come.

Two of Williams's intellectual disciples were Geoff Parker, then at Bristol University, and Robert (Bob) Trivers at Harvard, and I was lucky enough to be told about their research – some of it even before it was published – as an undergraduate. Parker studied dung flies and showed that male flies compete to inseminate females as they arrive at cowpats to lay their eggs. Contrary to what Darwin said, and what everyone therefore believed, female flies were not monogamous, and routinely mated with more than one male. Parker found that once a male has mated with a female he tries literally to hang on to her until she has laid her eggs, for only then will he be sure in his tiny pre-programmed dung-fly brain that those eggs have been fertilized by his sperm and not by those of a rival. Even if she has previously mated, it doesn't matter because, as Parker's clever experiments showed, the last male to mate before the female lays her eggs fathers most of the babies. Holding on to the female after mating was what Parker called 'post-copulatory mate guarding'. Trivers had noted the Feral Pigeons on the ledge outside his office window also engaged in mate guarding, with the male positioning himself between his mate and any other male as they went to roost.[15] In terms of genetic descendants, selfishness paid off.

Hearing about the research of Parker and Trivers as an undergraduate marked a turning point in my life. Completely seduced by the idea of individual selection (and sex), I'm sure it was partly as a result of my subsequently talking about natural selection in this way during my interview with David Lack that I secured my PhD placement. Trivers and Parker were two of the forerunners in a massive change in thinking partly centred on Oxford, where it was enthusiastically spearheaded by Richard Dawkins and John Krebs. It was my immense good fortune that my PhD coincided with the birth of behavioural ecology there.[16]

Inevitably, I suppose, this new approach to bird behaviour got off to a shaky start, mainly because the core idea of individual selection remained obstinately mired in its muddy past. It needed a book – something substantial – to kick-start it. Jerram Brown, an American ornithologist who studied cooperatively breeding jays, produced the forerunner in 1975, *The Evolution of Behavior*, which should have done the trick, but it was eclipsed by Edward O. Wilson's hugely impressive *Sociobiology*, published later the same year. Wilson was an expert on social insects and his book generated extraordinary publicity, in part because he discussed the evolution – that is, the genetic bases – of human behaviour, an area that had been off limits since the Second World War and the horrors of the Nazi regime. Left-wing biologists saw Wilson's book as justifying right-wing political policies, something that Wilson vehemently denied, sparking a controversy that was put to rest only in 2000 with the publication of Finnish sociologist Ullica Segerstråle's blockbuster *Defenders of the Truth*. Politics aside, *Sociobiology* set animal biologists thinking about new ways of studying behaviour. There was still a problem, however, because Wilson failed to be explicit about how selection worked.[17]

Dawkins to the rescue. Modestly produced in 1976, *The Selfish Gene* spelled out in elegant, lucid prose the wonder of individual selection. Using the gene as a focus was a stroke of genius, and the idea that a gene might be 'selfish' caught the public imagination. The book quickly became a bestseller and was – and still is – essential reading for every biology undergraduate. It took rather longer for this new way of thinking to penetrate the media, and throughout the 1980s and 1990s presenters of wildlife television documentaries continued to misinform the public by peddling Wynne-Edwards's ideas about why animals behaved in certain ways – for the good of the species – with Sir David Attenborough one of the few exceptions.

Three things made behavioural ecology a success: predictions, generalities and birds. Individual-selection thinking provided a theoretical superstructure for the subject, a framework that Tinbergen's and Lorenz's ethology never had. As a student, this was forcibly brought home to me during a brief conversation with John Krebs. With my interest in guillemots, I asked him why the seabird biologist Bryan Nelson, famous (to my mind at least) for his studies of Northern Gannet behaviour, was not part of the new movement. 'Because he never tested theory,' Krebs replied, and in an instant my thoughts were clarified. I knew then what I had to do: generate a hypothesis (an idea) and from that come up with some predictions, and find an elegant way to test them. On the basis of those tests, you will have an answer: Yes, No; True, False. Behavioural ecology was now *real* science, and we were using what is often called 'the scientific method' to obtain a better understanding of birds and the natural world.[18]

Female animals from species as diverse as insects, reptiles and birds are all likely to mate with more than one male during a breeding season. As a result they all experience sperm competition, and the way males protect their paternity or seek to make their own inseminations more effective have many similarities. The discovery of such 'generality' was important, for suddenly it wasn't sufficient to be interested only in birds. As a researcher one needed to read more widely, because what someone else had discovered in fish or reptiles or bats could inform and shape your ideas about birds.[19]

A lot of birdwatchers found an academic career studying the behavioural ecology of birds. Some scientists who had not been birders similarly discovered the utility of birds as subjects for testing ideas in behavioural ecology. Abundant, diverse, beautiful and, above all, easily visible, birds were the perfect animal group for testing the new ideas. At one point bird studies were so productive, informative and abundant that a senior journal editor, concerned by this taxonomic imbalance, announced that he would no longer accept papers on birds. It wasn't a problem, for there were plenty of other journals happy to accept high-quality ornithological papers, and our knowledge of birds, spurred on by the behavioural ecology revolution, continued its forward march.[20]

Although this revolution in evolutionary thinking made the study of birds particularly respectable, old habits died hard in some quarters. When I first started as a lecturer at Sheffield, the Zoology Department consisted entirely of endocrinologists, who referred to my field research as 'bird-watching' (not even 'birdwatching' – sensu Cocker) with both disbelief and condescension. There was similar bewilderment in my wife's family, with her mum struggling to understand why I was studying Eurasian Magpies. 'Well,' she said, 'I can watch the magpies in the garden. What makes what you do any different?' Explaining that I was collecting data to test a hypothesis sounded pretentious and condescending, so I dodged the issue, but it was an important lesson since it brought home to me that I should not take for granted that the purpose of my studies was easy to comprehend.

Several aspects of bird biology – notably, brood parasitism, cooperative breeding and mating systems – provided the perfect testing ground for behavioural ecology because, on the face of it, they seemed to challenge the idea that selection operated on individuals.

The Common Cuckoo's habit of duping other species into rearing its offspring had repulsed and puzzled people for centuries. Even after Darwin introduced the idea of natural selection, it still made no sense to most people that the cuckoo should reproduce successfully at the expense of its various host species such as the Meadow Pipit. Remarkably, even though it took biologists until the 1970s to fully understand natural selection, Darwin himself recognized perfectly

Common Cuckoo painted by Jemima Blackburn sometime between 1860 and 1880 (courtesy Alan Blackburn).

well in the 1850s how brood parasitism could evolve. Derived from first – that is, evolutionary – principles, Darwin's ideas were effectively hypotheses, hypotheses that Nick Davies set about testing at Cambridge in the 1980s. The elegance and efficacy with which Davies and others such as Steve Rothstein in the United States eventually confirmed Darwin's ideas, subsequently spawned an entire sub-discipline. We now know that brood parasitism has evolved several different times and that different species have evolved different ways of being a brood parasite.

The recently hatched Common Cuckoo ejects the host's eggs or chicks from the nest and as a result monopolizes its foster parents' efforts. But some species are crueller to their hosts. The young Greater Honeyguide inside the totally dark nest tunnel of its Little Bee-eater host emerges from the egg armed with a vicious spike on the tip of its upper mandible with which it kills its nest mates. I was lucky to spend time with Claire Spottiswoode in Zambia during her honeyguide research, in which she used tiny infrared cameras to witness the way the hosts' chicks were dispatched. One of the main revelations from all these studies is the ongoing arms race that exists between brood parasites and their hosts. No sooner has a brood parasite evolved a strategy for improving its chances than there is strong selection on hosts to evolve a counter-strategy to minimize or negate its effects.[21]

I previously mentioned a Note published in *British Birds* in 1907 about four adult Long-tailed Tits attending the young in a nest. It was probably assumed at the time that this was an aberration. In fact, it was an early record of cooperative breeding, whereby individuals other than the biological parents help to rear a brood. Such apparently selfless behaviour on the part of the 'helpers' seems to fly in the face of natural selection, and Darwin himself struggled to account for it, albeit in bees, wasps and ants, where the worker caste forgoes breeding to help their nest mates. The answer was one of the major triumphs of selfish-gene thinking, a decade before Dawkins wrote his influential book. This was the realization by Bill Hamilton that offspring share genes not only with their parents, but also with other relatives (their kin), and that by helping relatives, even without breeding themselves, individuals can pass on their genes to future generations. Hamilton's original objective was to explain altruism in social insects, but kin selection also accounted for the cooperative breeding in birds like Long-tailed Tits, Australian fairywrens, Arabian Babblers and Acorn Woodpeckers.[22]

Another challenge to selfish-gene thinking was deviation from monogamy. In a system operating for the good of the species, an equal sex ratio such that every individual might secure a partner seemed intuitively obvious. But not all animals, and not all birds, pair up in monogamous partnerships. Some, like the Long-tailed Widowbird, Black Grouse, Ruff, and Great Bustard are polygynous, one male mating with several females, while other species, including the Spotted Sandpiper and phalaropes, are polyandrous, one female mating with several males. These different mating systems required a selfish-gene type explanation, and from the mid-1970s became the focus of research among ornithologists. The answer was that different ecologies allowed certain males or females to monopolize members of the opposite sex to maximize their own reproductive success.

Because most birds seemed to be monogamous, like ourselves, monogamy was considered rather boring and not very attractive as a topic for research. As it turned out, monogamy proved to be much more fascinating than initially assumed.[23]

Inspired by Geoff Parker's dung-fly studies, I wondered as an undergraduate whether promiscuity and mate guarding also occurred in birds. I was told I was wasting my time because 'everyone knew' that birds were monogamous – Darwin himself had said so, as had David Lack, my mentor at Oxford. Reading as many accounts of bird behaviour as I could find, I discovered remnants of a prudish Victorian legacy in which details of anything to do with mating were studiously ignored in what were otherwise commendably detailed accounts. Clearly, the sexual revolution of the 1960s had yet to penetrate ornithology's hallowed walls. Nonetheless, I found encouraging hints, including Edmund Selous's observations of Common Blackbirds made in 1901: 'The male bird follows her all about, hopping where she hops . . . the cock is busy escorting and observing the hen.' Perfect! I knew I was on to something. All I had to do was to identify a suitable study species – I chose the Eurasian Magpie – and collect the necessary data mapping the male's proximity to his partner in relation to her fertility.[24]

At that time no one knew precisely when female birds were fertile, but it was most likely to be in the few days before and during egg laying, and that's precisely when I found the male magpie closest to his partner. And what were males actually guarding against? I had assumed that, as in the dung flies, it was from the sexual advances of other males. Frustratingly, I saw precious little evidence for this as I sat hour after hour, day after day, peering through my telescope at the magpies. I consoled myself by thinking that guarding was like an insurance policy and was not needed most of the time. And then the day came when the value if not the efficacy of magpie mate guarding was laid out in front of me more dramatically than I could ever have hoped.

I had been watching a particular pair over several days as they built their nest, with the male trailing round relentlessly after his partner, walking where she walked, flying when she flew, hour after hour. It had been a cold, blustery spring and this was the first day of warm sunshine. The male sat on a drystone wall watching his partner as she foraged just a few metres away. As I observed them, I saw to my amazement that he had fallen asleep, with his head now below the level of the wall. No sooner had I realized he could no longer see his partner than the male from the adjacent territory noticed too, and in a flash flew over and mounted 'my' female. I watched in disbelief. The sleeping male awoke, the intruder rapidly withdrew, and the male chased his partner back to the nest, as though to say, 'Enough of that!' But the deed was done – the deed being a mating outside magpie marriage – and may have resulted in one or more extra-pair offspring. At that time there was no way of knowing if the intruder had been successful, for molecular paternity tests were some way off. But witnessing such a blatant example of cuckoldry right when the female was fertile was enough, and provided the glorious icing on my mate guarding cake.[25]

The behavioural ecology revolution created hundreds of jobs for ornithologists. Birds were among the most popular of animals for study, and all universities wanted to be able to offer courses in behavioural ecology. Many undergraduates arriving at university as birdwatchers were transformed into scientists, attracted by the evolutionary logic provided by this new way of thinking. It was a methodology that allowed them to test ideas in a way that had never been possible for the ethologists in the first half of the twentieth century.

All this, of course, resulted in a massive increase in our understanding and appreciation of bird behaviour. In 1960 the number of papers published on birds each year was around 2,500, but by 2020 this had increased to almost 20,000. Behavioural ecology had arrived. By the 1990s, the spotlight was no longer shining solely on Tinbergen's adaptive significance question, but now, like a searchlight arcing across the night skies, it was embracing his other three questions as well in a quest for more comprehensive answers. In my own research on sperm competition in birds I went on to explore the physiological processes that determine which of a female's several partners fertilizes her eggs.[26]

From the 1990s until the present day, bird researchers have continued to broaden their outlook. Still using an evolutionary framework, they have gained better insights and, in some cases, true understanding of a huge range of topics that span community ecology, ecosystem ecology, systematics, comparative anatomy, physiology and conservation biology. Such has been the success of behavioural ecology – and the selfish-gene way of thinking – that fifty years after its inception these remain vibrant, fast-moving areas of bird research. The take-home message of birding (Chapter 9) and the biology we have discussed in this chapter – as well as conservation biology in the next chapter – is that our understanding of birds and, crucially, our relationship with them, continues to evolve.

12. Ghost of the Great Auk: Third Mass Extinction

I would write the story of the biblical Samaritan again. He's someone who feels we are living in the last days of human restraint, and in the last days, too, of the Northern Bald Ibis of the Ethiopian highlands and the Spoon-billed Sandpiper of the Chukotski peninsula . . . but in his own small way he decides to help, not turn his back . . .

Barry Lopez (2019)

Alfred Newton – pedant, procrastinator, perfectionist, misogynist, bibliophile, conservative and snob – was the greatest of all Victorian ornithologists. When he died of congestive heart failure at the age of eighty-seven on 7 June 1907, a career dominated by the study of death and extinction came to an end. Newton's eminence as an ornithologist flowed from his encyclopedic knowledge of birds, from his position as first Professor of Zoology and Comparative Anatomy at Cambridge and one of the founders of the British Ornithologists' Union (and editor of their journal *The Ibis*), and as the brains behind the monumental *Dictionary of Birds* published in 1896.[1]

A typical Oxbridge don, Newton never married and was cared for by his college, Magdalene, where he lived and died. He was learned, enthusiastic and hugely encouraging to any undergraduates (all male at that time) expressing even the slightest interest in birds. In an era with several larger-than-life ornithologists, Newton was the most influential.[2] Yet for all his sparkling scholarship, Newton made no real scientific discoveries. Rather, he was an accumulator of books, of facts, of eggs and museum skins, and was deeply averse to any kind of theorizing or speculation. But what he did do – and for which he should be hailed a hero – was to launch the enterprise of bird protection.

By the mid-1840s, the engineering miracle that was the British railway network meant that people could travel from any major city in Britain to coastal resorts like Scarborough and Bridlington, where the bracing sea air and chilly waters were said to promote health and happiness. From Bridlington one could hire a small boat to take you to the foot of the 400ft chalk cliffs at Bempton on the Flamborough headland to shoot Black-legged Kittiwakes – on the sea, on their nests or in the air – to your heart's content. Here, happiness was a warm gun. The seabirds on this headland were hit hard in Victorian times, first by the farm labourers known as 'climmers', who descended on ropes to harvest vast numbers of Common Guillemot and Razorbill eggs each year, but also by those armed with shotguns, advancing through the waves: 'Parties of sportsmen, from all parts of the kingdom, visit Flamborough . . . and spread sad devastation all around them. No profit attends the carnage; the poor unfortunate birds serve merely as marks to aim at, and they are generally left where they fall.'[3]

Addressing a meeting of the British Association for the Advancement of Science in Norwich in 1868, Newton reminded his audience that while there was long-standing legislation protecting herons and game birds (so that 'sportsmen' could kill them), seabirds, raptors and common garden birds had no such protection: 'With reference to seafowl a certain amount of sentiment may be confessed. No animals are so cruelly persecuted. At the breeding season they come to our shores, throwing aside their wary and suspicious habits . . . Each bird shot was a parent, and its young were thus exposed to death from hunger.'[4]

Newton told his audience how one shooter at Bempton claimed to have killed 4,000 kittiwakes in a single season, and of another who had received an order from a London millinery house, for 10,000 of these small gulls. Newton's call for a 'close season' received widespread public support, albeit somewhat offset by his careless implication that those responsible for the slaughter were the local people of Bridlington rather than outsiders.

Newton's passion for protecting birds had been brewing for some time. In 1858, at the age of twenty-nine, he and his friend John Wolley had set off for Iceland in the hope of finding a few Great Auks still alive. They were unsuccessful. The last two individuals had been killed in 1844 on the island of Eldey, about sixteen kilometres off the south-west tip of Iceland. Frustrated by not being able to get to Eldey because of heavy seas, Newton and Wolley decided to interview some of the men involved in the 1844 trip. In so doing they obtained an unprecedented record of the precise moment in which a species fell irrevocably from extant to extinct. Newton was captivated by the Great Auk, or Garefowl, as he preferred to call it, and for the next twenty-five years continued to collect information on it and its very scarce eggs. An arch-procrastinator, his plans to produce the definitive account of this iconic and charismatic seabird were dashed when, in 1885, he was scooped by the relatively unknown Symington Grieve.[5]

Those two months in Iceland, haunted by the ghost of the Great Auk, shaped much of Newton's subsequent career. Most important was his realization that extinction was a topic open to scientific investigation, for at that time it was almost inconceivable that an entire species like the Great Auk, once so abundant, could be totally eliminated by the hands of men. Yet, sixty years earlier in 1785, the adventurer George Cartwright, appalled by the relentless commercial slaughter of Great Auks on Funk Island, off the north-eastern Newfoundland coast, commented that unless the killing ceased, 'the whole breed will be diminished to almost nothing'. This is exactly what happened because as the bird became increasingly scarce, the market price for eggs and skins soared . . . and the killing continued.[6]

Newton's interests extended to other extinct birds, including the Dodo, the Solitaire and the Crested Parrot, all once common on the Mascarene archipelago in the Indian Ocean, 700–1,000 kilometres east of Madagascar, as well as the local extinction of the Great Bustard in Britain. Somewhat paradoxically, given his conservatism, Newton's scientific

A male Great Auk accompanies its chick to the sea at fledging (drawing by David Quinn).

interest in extinction was reinforced by the publication of Darwin's *Origin* and the recognition that extinction was an inevitable consequence of the long-term dynamics of natural selection.

It was the contrast between *natural* extinctions, like that of *Archaeopteryx*, and *unnatural* extinctions, like that of the Great Auk caused by people, that ignited Newton's passion for bird protection. As Newton saw it, a species that becomes extinct through natural causes is replaced by another, better adapted species. The tragedy of unnatural extinctions, he realized, was that those species are not replaced – their loss remains an unfilled gap among the branches of the tree of life.

Newton's sentimental sense of indignation at unnatural extirpations was rooted in science because he felt that such losses deprived people from learning about those species and made it even harder to understand the workings of nature. At a time – the mid-1800s – when many Victorians were agitating about unnecessary cruelty to animals, Newton found himself in a dilemma over his concern for persecuted birds. He justified his ideas by dividing sentimentalists into, on the one hand, those who were 'opposed to the killing of birds for almost any reason' and, on the other, those 'who acknowledged the dominion of man over nature, but asked only that it not be abused'. This latter category included ornithologists who killed birds or took eggs for science. But there was still a contradiction, for it was scientists – and their bird collectors – who finally extinguished the Great Auk.

Sentimentality towards animals was deemed both 'unscientific' and unmanly, a controversy that Darwin himself was caught up in. A self-avowed animal lover, Darwin abhorred cruelty, but also acknowledged that some experiments on live animals, usually dogs, were necessary to further our understanding of physiology and medicine. Writing to a zoologist colleague, he said: 'You ask about my opinion on vivisection. I quite agree that it is justifiable for real investigations on physiology, but not for mere damnable and detestable curiosity.' Darwin's views were naturally derived from his belief that humans and non-humans are all animals.[7]

It was the shocking realization that humans could cause extinction that motivated Newton's protectionist ideas, forcing him to adopt a more sympathetic approach than he would usually have felt comfortable with. The difference was that those he considered to be extreme sentimentalists focused their concerns on cruelty inflicted on *individual* animals, whereas his concern was with the global or local populations of entire species.

Newton's ideas mark a turning point, 'a particular moment in the history of Victorian Britain and in the history of science . . . in which the boundaries between science and sentiment, and between those who did and those who did not have the authority to speak for nature, were being redrawn'.[8] His research on extinct and near-extinct birds gave Newton a unique understanding of the process distinct from that of geologists and palaeontologists, whose knowledge of extinction was confined to species long-since departed. Newton's understanding of *how* extirpation happened placed him – and other naturalists – in the unprecedented position of acting as advisers on the practicalities of protection and, crucially, as the creators of a set of guidelines for a more general animal protection movement. Up until this time, legislation for protecting game species, usually from poachers, had come from landowners. The time for naturalists to take the lead had arrived.

In response to Newton's British Association address, Henry Frederick Barnes-Lawrence, perpetual curate of Bridlington Priory, called a meeting of local clergymen and naturalists. His aim was partly to pacify the Bridlington residents who had been falsely accused of harming the local seabirds, but primarily to bring an end to the killing of seabirds on the Flamborough headland. One of his most enthusiastic supporters was a well-known local bird author, the Reverend Francis Orpen Morris, who, in a letter to *The Times* newspaper described how 'hosts of excursionists from the town of Sheffield, which has lately obtained an unenviable notoriety for a slaughterous propensity [to] wage cowardly and murderous war on the defenceless and timid birds.'[9]

Barnes-Lawrence had been convinced of the seriousness of the problem by naval commander and amateur zoologist Hugh Horatio Knocker, whose calculations revealed that 'roughies' were responsible for killing no fewer than 100,000 seabirds at Flamborough each breeding season.[10]

The result was the formation of the Association for the Protection of Seabirds, supported by the great and the good, including Morris himself, the naturalist and writer Frank Buckland, the solicitor Thomas Harland – known as the 'father of Bridlington' for his public works – and the local Conservative MP, Christopher Sykes, who was charged with introducing the Bill to Parliament. Poor Sykes, one cannot help but feel sorry for him. Considered 'a shocking snob' and the butt of endless jokes, his speech in support of

'Roughies' shooting seabirds at Bempton, Yorkshire, in the early 1900s (courtesy Mrs M. Traves).

seabirds on 6 March 1869 was the first time he had spoken in the House of Commons since his election four years earlier. Following the petition, he was portrayed in *Spitting Image* style in the magazine *Vanity Fair* as 'the gull's friend' – an allusion to his gullibility. Despite Sykes's dubious status, the Bill imposing a closed season on the killing of seabirds passed, and later that year received Royal Assent.

But bird protection was not easily legalized, and over the next eleven years Newton campaigned tirelessly, writing dozens of letters to everyone who might have influence. It is not surprising that the enactment of this legislation was so difficult, for apart from the rights and wrongs of shooting, the Bill was riddled with class snobbery and contradictions. As Francis Morris snobbishly and unashamedly revealed in another letter to *The Times*, the Bill was directed specifically at working-class shooters: 'sportsmen of course they are not'.

One of the Bill's strongest supporters, Prince Christian, Queen Victoria's son-in-law and therefore sufficiently above the law, saw no contradiction in asking Barnes-Lawrence to send some Bempton seabird eggs to him at Balmoral. Even Alfred Newton argued that egg collecting be excluded from the Bill because 'the eggs of many seabirds are taken by country people for food', because many 'poor families . . . earn an honest living . . . by collecting [seabird] eggs', and because some local people, like those on the remote island of St Kilda, depended almost entirely on seabirds. Oh yes, and because Newton himself collected eggs for his museum.[11]

Not everyone approved of the new law: 'a fowler' wrote to the *Yorkshire Gazette* asking what use or ornament seabirds were to people, if they weren't there to be shot at? But the Act had teeth, for on 10 July 1869 Barnes-Lawrence noted with satisfaction in his diary how a Mr Tasker of Sheffield, who openly shot twenty-eight birds in defiance of the law, was apprehended and fined 2/6 for each bird and 9/- costs: a total of £3 19s. 'Thank God for that,' Barnes-Lawrence sighed.[12]

What got the Seabird Protection Act through Parliament was that it was less about preserving seabirds than about protecting people's lives and livelihoods. The cries of seabirds at their breeding colonies, it was argued, warned fog-bound mariners of the proximity of rocky shores, aggregations of seabirds at sea showed fishermen where to cast their nets and gulls following the plough helped farmers by devouring innumerable noxious invertebrates.

A year after the Act was passed, a pronounced increase in seabird numbers was noted, and Newton claimed that the Act had saved the guillemot from extinction – at Flamborough, at least. Nevertheless, shooters continued to be prosecuted for killing kittiwakes there for several more years.[13]

The fact that seabird protection saw its beginnings at Bempton creates a warm glow in my heart, for I have strong links with this historic colony and with the conservation of seabirds. Bempton is the closest and most accessible seabird site to where I grew up (and to where I now live in Sheffield), and during my PhD studies I spent many a winter's dawn there, when my main study site, Skomer Island, was inaccessible. As a result of protection, the numbers of Common Guillemots and other seabirds have increased considerably since I first visited Bempton in 1973. It then felt very remote, for this was long before there was a Seabird Centre there, and outside the breeding season there were few visitors. Arriving in the dark, I hunkered down on the cliff tops, close to the edge, wrapped in multiple layers, waiting for the birds to appear. The only sounds were the wind and the waves crashing on the chalk boulders below. As the first streaks of light appeared on the eastern horizon, the guillemots arrived en masse from the sea, announcing their arrival with an exhilarating chorus of greeting calls as partners were reunited. With the light increasing and with my brain flooded with endorphins, I watched and noted the guillemots' behaviour as they reaffirmed their pair bonds with caresses and copulations.

Despite its inadequacies, the Seabird Protection Act of 1869 was a start, and one in which Alfred Newton played a key role. The Act was an ornithological milestone marking the culmination of a long campaign dating back to the 1640s for the better treatment of animals, but also a fingerpost anticipating future conservation efforts.[14]

Changing people's opinions about animal cruelty was a difficult and protracted business. The first attempts to curb bull-baiting and cock-fighting in England in the early 1800s, for example, were met with disbelief and derision in Parliament:

'the next step will be to protect flies and beetles', opponents said.[15] Every attempt to protect animals seemed to be met by resistance and ridicule. Eventually, though, a group of enthusiasts formed a Society for the Prevention of Cruelty to Animals (SPCA) in 1824 and were given royal approval by Queen Victoria in 1840. Royal patronage was almost essential, for, as we have seen, animal welfare was very much a class issue, championed by the intelligentsia – writers like Thomas Hardy and John Ruskin – and the aristocracy, for as well as promoting animal welfare, their mission was to suppress the dangerous elements in human society. As Harriet Ritvo says in her *Animal Estate*, identifying cruelty as 'a lower-class propensity' created a 'moral distinction which [the RSPCA's] potential patrons found comforting and attractive'. The society's objectives, spelled out at their first meeting in 1824, were: not only 'to prevent . . . cruelty towards animals, but to spread among the lower orders of the people . . . a degree of moral feeling which would compel them to think and act like those of a superior class'.[16] One way the RSPCA did this was to appeal to people's emotions by portraying 'suffering animals', mainly domestic species, as 'noble, selfless' servants (think of Anna Sewell's *Black Beauty*), and the abusers as 'rough members of the urban proletariat'.[17] Appealing to people's emotions was what helped to prevent the slaughter of innocent seabirds at Flamborough in the 1860s and has played a crucial role in all subsequent bird protection.

Some of the Black-legged Kittiwakes shot at Bempton were scooped up from the sea and taken to Bridlington, where their wings and breast feathers were removed and cleaned.

With a suitably large haul, these were then shipped off to London fashion houses, where plumassiers converted them into elaborate and expensive headgear for ladies.[18]

The wings, tails and plumes of birds such as kittiwakes, and more exotic species like egrets, birds of paradise, tanagers and hummingbirds, incorporated into hats became synonymous with Edwardian fashion. The marketing line was that such millinery would enhance the wearer's status or

The 'Extinction of the species' caused by the fashion for feathers (from *Punch Magazine*, 6 September 1899).

attractiveness – just as it did for the birds that originally owned them, of course (although this was never explicitly stated) – and as other forms of feather-wear had done for the indigenous peoples of the Americas, the South Pacific and Africa. For the feather-pluckers and fashion houses of London, Paris and New York, feathers were immensely lucrative.

Disgusted by the cruelty and monumental loss of bird life occasioned by the feather trade, Emily Williamson, a resident of the Lancashire town of Didsbury, in England's industrial heartland, and her friend Etta Lemon of Reigate in Surrey, established the Society for the Protection of Birds in 1889. This was a group that initially consisted entirely of women – and mostly women of wealth – whose aim was to bring the wearing of 'murderous millinery' to an end. Appropriately, the fledgling society's first guinea of support came from Alfred Newton. Another male ally was the writer and bird lover William Henry Hudson. As an anti-Darwinian and arch-sentimentalist, he could hardly have been more different from Newton, but Hudson was exactly what the society needed: a wordsmith to pen their publicity pamphlets. The society was so popular and effective that in 1904 it received its royal charter, becoming the Royal Society for the Protection of Birds (RSPB). Nonetheless, it took until 1921 for the Importation of Plumage Act to pass into law.

In North America, in an almost identical scenario, three wealthy, well-connected women, Florence Merriam Bailey, Harriet Hemenway and her cousin Minna Hall, encouraged others to boycott the feather trade, and in 1896 Hemenway and Hall founded the Massachusetts National Audubon Society with the aim of publicizing bird protection. The Migratory Bird Treaty Act of 1918 put an end to the plume trade. North America was ahead of Britain in protecting birds thanks to the naturalists John Muir and Henry David Thoreau, who in the 1850s conspicuously discarded the bird-killing mantle that had made John James Audubon famous – or infamous. Presciently, as early as 1853, Thoreau had wondered, 'Would it not be well to carry a spy-glass in order to watch . . . shy birds such as ducks and hawks?'[19] Even the scientific American Ornithologists' Union, under the leadership of J. A. Allen, set up a committee in 1886 to ensure the legal protection of birds. I say 'even', because it took the British Ornithologists' Union another sixty years to shake off the Victorian 'leprosy of collecting'.[20]

As the killing of birds for their feathers became illegal, the killing of birds for fun was allowed to continue, at least for those species designated as 'game' and worth eating, such as partridges, pheasants, grouse, waterfowl and wading birds. Also allowed to continue was the collecting of birds and their eggs for science. For those birds destined for the dinner table and those destined for museum cabinets, the numbers involved were far lower than the wholesale slaughter resulting from the plume trade. Both the aristocracy and scientists were outside the law; the former because they made the law, the scientists – collectors like Allan Hume – because their endeavours were considered worthy.

The first few decades of the twentieth century saw attitudes towards birds changing. Shooting 'game birds' continued, and continues to this day, albeit increasingly controversially, but the scientific collecting of skins and eggs started to shrivel, replaced – as we saw in the previous chapter – by the start of birdwatching and an increasing interest in how birds live.

By the 1920s Max Nicholson recognized that the first step in protecting birds depended upon knowing how many there were. To this end he launched a number of important national surveys. The idea was not new, for back in 1861 Alfred Newton had suggested a national census of birds. It was a notion that came from his friend John Wolley, who suggested that this was 'the chief requirement of British ornithology'. It was only after Wolley died in 1859, that Newton put pen to paper, outlining the pros and cons of a national bird census. He struggled to see how such a census could be accomplished, but its benefits, he said, were clear. In the years immediately following the publication of Darwin's *Origin* – with expressions such as the 'struggle for life' ringing loudly in people's ears – Newton knew that knowledge of bird abundance would provide a better understanding of such fundamental issues. Newton was prescient in recognizing that any national census could be undertaken only with the 'cooperation of nearly all the ornithologists in the country'.[21]

The first of Nicholson's surveys, conducted in 1928, was a census of heronries, an appraisal of Grey Heron numbers in England and Wales, undertaken by no fewer than 400 volunteers. That census continues to this day. My own attraction to this species in the 1960s, and to the value of censusing, was partly due to the heron's iconic status as the first bird to be subject to long-term monitoring. In 1928 the heron was not in need of protection, but Nicholson knew exactly what he was doing. The heron's conspicuous treetop nests allowed its numbers to be easily assessed, and it therefore served as a valuable test case for monitoring projects on other species. Following hard on the heron survey's heels, another census, undertaken in 1931, also by volunteers, focused on the Great Crested Grebe, whose population had been plundered by the plume trade. This was a survey inspired by Nicholson and, after an intimidating interview with him, organized by Phil Hollom and Tom Harrisson, both barely out of their teens and both of whom went on to great ornithological achievements. Hollom – 'Gentle Phil', as he was known, and an Establishment figure – became one of Britain's best-known birders and co-author in 1954 of one of the first field guides to British birds. The precociously clever and staggeringly arrogant Harrisson – 'the most offending soul on earth' – went on to excel as an ornithologist, journalist, soldier, archaeologist, writer and conservationist. Max Nicholson himself went on to extraordinary achievements in conservation, including the founding of the World Wildlife Fund (WWF) in 1961. The Great Crested Grebe survey, with its 1,300 volunteers, presaged a future enthusiasm for similar surveys of many more species, and just as Alfred Newton anticipated, these surveys were driven by amateurs. Establishing ways of accurately and reliably monitoring the numbers of different bird species took some time, but since the late 1960s the British Trust for Ornithology in Britain and the Cornell Laboratory of Ornithology in the USA have organized and motivated networks of volunteers whose efforts have provided a clear view of the status of these countries' bird populations. The results, mirrored across the world, show an ongoing and apparently unstoppable decline in the numbers of many bird species.[22]

I grew up with Nicholson's mantra of 'useful' birding ringing loudly in my ears. As a teenager I drove many miles in my mum's Mini, counting rookeries and heronries and even publishing the results of those surveys. They provided instant gratification through the sense of having achieved something useful. In contrast, studying my other passion, bird behaviour, seemed about as tractable as removing an angry Blue Tit from a mist-net – until someone shows you the way.

Two events around the age of eleven helped to direct my ornithological future. The first was when my mother's uncle, my Great Uncle Roy, a keen birdwatcher, presented me with a special notebook with my name and the words 'Bird Notes' embossed on the cover, in which to describe and illustrate my observations. This simple, thoughtful gift elevated my birding activities into something indefinably significant, and, as I now realize, helped to fuel my ambition. The second event occurred on a single, halcyon summer's day when, during a holiday in North Wales, my father took me to Bardsey Island (Ynys Enlli), off the tip of the Lleyn peninsula. I had never been anywhere quite so perfect and it was here, amidst Northern Ravens, Red-billed Choughs, Razorbills and Common Guillemots, that the fantasy of living on islands took root. As my father and I walked back towards the boat at the end of the day, we passed a young man watching seabirds through a telescope with a field notebook open in his lap. After acknowledging him, my father turned to me as we walked away and said, 'You could do something like that when you are older.'

During my second year as an undergraduate at Newcastle, I saw pinned to the departmental noticeboard details of a bird conference especially for undergraduate students, organized by the Edward Grey Institute in Oxford. My birding buddy Rob Taylor and I went together, and for several days sat on hard seats in St Hugh's College, awed by the institute's director, David Lack, and his various academic colleagues and listened to students talk about their studies. Naively, for I'd never spoken in public before, I had offered to speak about my heron project. Even though I felt my talk was dreadful, Lack nonetheless offered me a PhD place – to study Common Guillemots on the Welsh island of Skomer. I couldn't believe my luck, and it suddenly seemed as though the diverse loose threads of my life were now weaving themselves into something substantial.

The number of guillemots – along with those of the Atlantic Puffin and Razorbill, all close relatives of the now extinct Great Auk – had apparently been declining in southern Britain and no one knew why. The aim of my PhD was to find the answer. I was part of a trio of graduate students: Clare Lloyd on the neighbouring island of Skokholm was studying Razorbills and Ruth Ashcroft was to do the same for the Atlantic Puffin with me on Skomer.

David Lack was the acknowledged authority on bird populations and the author of several acclaimed books, so I seemed to be in the right place.

My studies on Skomer started in 1972, the year I graduated, and my PhD ran from 1973 to 1975. At that time, the heart-rending sight of guillemots and other seabirds fatally soiled by the crude oil washed or spilled from the bellies of giant tankers was all too common. Appalled by the idea that auks on Skomer and elsewhere were declining, I was keen to do whatever I could to help, and understanding their population biology was a start. At the same time, the more I watched 'my' guillemots, the more enthralled I became by the complexities and subtleties of their social lives. Less cute, less colourful and less charismatic than puffins, there is something especially engaging about the intricate web of adaptations that allows a chicken-sized bird to lay and incubate its single, pear-shaped egg in the open on precarious cliff ledges amidst a throng of noisy neighbours.

Studying both population biology and behaviour stood me in good stead, since in the 1970s conservation was not considered academically respectable. Conservation was important, but the authorities then believed that bird preservation should be funded by charities rather than by the government's scientific research councils. In other words, if you wanted an academic career you focused on 'serious science', in this case behavioural ecology, so that is what I did. At the same time I continued to think about how my work might contribute towards conservation. Luckily, over the following decades it became clear to the authorities that, to be effective, conservation required 'serious science'. Understanding the way populations worked, however, *was* respectable, but I quickly realized that getting results from a bird that was as long-lived as the guillemot was going to take too long to yield the mass of published scientific data that were (and still are) so essential for getting more research grants, for building the sort of academic empire the universities wanted, and for my personal promotion.

As Nicholson had rightly stated in the 1920s, the most fundamental aspect of studying the population dynamics of any species is to know how many there are. Luckily for me, Skomer's first warden, David Saunders, had made annual counts of all the seabirds on the island in the early 1960s, and it was these that confirmed the ongoing decline in guillemot numbers. Even though the population was relatively small at that time, counting Skomer's guillemots required a huge effort and a lot of queasy bobbing around with binoculars in boats below the cliffs. Something more efficient was required, and I identified some sample areas ('study plots') that could be counted more rapidly from land and whose changing numbers closely reflected the changes in the entire population. A second aspect was understanding just when during the four-month-long breeding season those counts should be made. I found the answer by counting the birds every day between April and July and finding the part of the breeding season that numbers on the cliffs were least variable – it was when the birds were incubating their eggs. Devising a consistent way of counting was the key to reliability.

The second part of studying how a population 'works' – in this case, why it was declining – involves knowing how long birds live and how many offspring they produce. If a population is to remain stable, those two have to balance each other in the long term. Longevity, determined ultimately by year-to-year 'survival', involves catching some birds and marking them with a ring whose unique number can be read from a distance without disturbing them. The proportion of birds surviving from one year to the next is our measure of survival, and in the guillemot's case this method works well because most birds that are still alive return to the same tiny breeding site on the cliff year after year. It sounds straightforward, but guillemots breed in such dense colonies that trying see their coloured leg rings is like trying to see the socks on the spectators in a crowded football stadium.

Catching guillemots was a challenge. It always meant scaling a cliff to reach a point where you could safely employ a Faroese-style *fleyg* or elongated crook to lift a bird carefully off the ledge.[23] The guillemot was once known as the 'foolish guillemot' in reference to the fact that, when approached by a predator such as a gull, raven or human, they sit tight on their egg or chick rather than flying away. For human predators such foolishness was a gift, since it made catching guillemots relatively easy. Of course, guillemots have only recently encountered human predators, and by breeding on remote islands or narrow cliff ledges they generally avoided terrestrial predators like foxes. Sitting tight in a dense group proved an excellent defence against gulls and ravens.

Uncomfortably trussed up in my climbing harness, and wearing a safety helmet, I am securely attached by a rope to a couple of pitons hammered firmly into the cliff face. In my hand is an ultra-light extendable fishing pole, to which I have added a miniature shepherd's crook at one end. I'm just a few metres above a ledge on which sit a mass of noisy guillemots that are not as foolish as was once thought, for if I get any closer they will leave. That's the last thing I want, and the purpose of the long pole is so that I can catch them with as little disturbance as possible.

I'm hot and tense from the exertion of the climb and sweat is running down from my helmet straps into my eyes. A quick wipe would help, but just getting here has covered my hands in guillemot shit and I do not want to risk the eye infection a colleague once experienced. I shake the sweat away and wait, looking carefully at the birds in front of me. They are tense too, but they sit tight. I have been catching guillemots on this ledge each year for over three decades so I kid myself that they are used to me: 'Oh God, here he is again!' Many of the birds bear rings from earlier years, so I'm looking specifically for those without rings. I see one and, careful to avoid any jerky movement, I slowly slide the tip of the pole towards it at foot level. It's a bit like hooking ducks at the fair. Despite its clunky appearance, if it decides to dodge my hook, a guillemot acquires the agility of one of Michael Flatley's dancers. The bird looks suspicious, so I pause. Then, by luck, there's a flurry of wings as another guillemot arrives nearby. Momentarily distracted, my bird looks up, and I gently slide the hook around its 'ankle'. It's on! And very slowly I start to walk the bird towards me. It seems mystified: 'By what hidden force am I being drawn to the monster on the ledge above me?' I take the bird with both hands, holding its wings against its body so it cannot flap, but this is a kilogram of pure muscle armed with a sharp beak so I have to be careful. Fortunately, unlike their Razorbill and Atlantic Puffin cousins, Common Guillemots are placid prisoners and almost never lash out. I place the bird under one arm with its feet facing forward. Reaching for my 'ringing bag' I take out the pliers and two rings: one metal, designed to last longer than any bird's lifetime, the other coloured plastic, bearing a unique engraved number that we can read through a telescope from a kilometre away. Both rings can be buggers to put on. The metal ring – it's called a 'guillemot special' – is designed so that its number cannot be erased by wear and tear inflicted by the rocks, but it requires a real knack to place it correctly on the bird's leg. I've been putting these on for decades and no longer give it a second thought, but the difficulty becomes all too apparent if I try to show someone new how to do it. The plastic rings are so tough I have to use circlip pliers to place them on the bird's leg, bracing the ring against my thumb in case the pliers slip. After several hundred rings, my thumb feels as though it has been through a mangle. Once a ring is on each leg, I call out the numbers to my field assistant, double check that they are correctly noted and then launch the bird into the air away from the cliff. Phew! And now on to the next bird: and on and on. I have to catch around fifty and I'm keen to do so as swiftly as possible to minimize the disturbance. As I search for my next bird, there's another whirr of wings and the one I have just ringed is back on its breeding ledge, casting me a disapproving glance as it settles back onto its egg.

I enjoy catching guillemots. It is tense and dirty work, but knowing that we might see each marked bird for the next ten, twenty or thirty years is an inspiration. I also love the feel of a guillemot in my hands, the velvet-like feathers on its head and its gentle dark-brown eyes. Our proximity to the birds, the cacophony of their calls and the smell, together with the camaraderie and teamwork of my three or four colleagues, make this an almost spiritual experience. In fact, the single day on which we catch and ring guillemots is the highlight of each year and one whose success I always celebrate with a team photograph.

While ringing, if we see any birds wearing only a metal ring we try to catch them in case it is one that has somehow lost its plastic counterpart. I once hooked such a bird and without looking at the ring, passed it back to my colleague, who then shouted: 'Effin' 'ell, it's a foreign ring!' And so it was. A guillemot ringed as a chick by a Swedish colleague on the island of Stora Karlsö in the Baltic, off the coast of Gotland, several years previously, that had decided to emigrate and breed on Skomer. A rare instance of immigration: magic!

At the end of the day my colleagues and I unclip our carabiners from the safety rope, pack up and (attached to another rope) climb back up the cliff, knowing that we have topped up our population of ringed birds from which we are able to estimate their survival. Our oldest birds are now in their late thirties, and on average about 95 per cent of them survive from one year to the next, which means that on average they have a breeding life of about twenty-five years.

The other variable in the population equation is what we sometimes call 'productivity', the number of offspring produced. Common Guillemots lay a single large, brightly coloured egg that is extremely pointed at one end, a shape that makes it more stable on the bare rock surface.[24] The bright colours and varied patterning of guillemot eggs have evolved to help parents to identify their own amidst those of their many neighbours. Despite the relatively large size of a guillemot egg (equivalent, relative to body size, to a 16lb/7kg human baby), they are remarkably difficult to see, partly because of the crowded conditions, but also because they are so closely covered by the incubating parent. We need to know how many pairs have laid an egg and, after the thirty-two-day incubation period, how many hatch. To do this we sit inside a tiny wooden hide perched on the opposite cliff and for eight hours each day scan the colony, hoping to confirm the presence or absence of an egg under each bird. Some guillemots lose their eggs through accidents, such as a fight between neighbours, or because the egg is taken by a predatory gull or raven. Once the egg hatches, a chick's presence is easier to record because it moves around and is fed by a parent several times each day. After three short weeks the flightless chick leaps from its ledge and swims out to sea accompanied by its father. Overall, around 80 per cent of our guillemot pairs produce a chick that survives to leave the colony.

Is this enough – a balance between productivity and survival – to keep the population stable? That depends on several other things, notably the age at which the young birds start to breed and the proportion of them that survive to that age of first breeding. Establishing how old they are when they first breed is a simple (but laborious) matter of looking for

Guillemots – a bridled bird in summer plumage (left), a winter-plumaged bird (centre) and an immature or winter-plumaged Razorbill (right), painted by Jemima Blackburn in the late 1800s (courtesy of Alan Blackburn).

birds ringed as chicks (and hence of known age) and recording when they first breed. This is generally at seven years old, with a few starting in their fourth and some unfortunates still without a mate at age ten. About half of our ringed chicks eventually return to the colony to breed. This is very high compared with most birds, where usually only a few per cent return, and it is this high survival rate of young guillemots together with the high survival rate of adult birds and their high rate of breeding success that account for Skomer's steadily increasing population. It sounds straightforward, but it took a full thirty years and thousands of hours of research to establish this.[25]

Our research reveals *how* the increase in population size has occurred, but it doesn't tell us *why*. The most obvious answer is that there is more food now than there was previously, but establishing the abundance of those fish that guillemots rely upon is extraordinarily difficult. The last few decades have seen less of the oil pollution that was previously one of the factors causing the population to plummet to its low level in the 1970s. Photographs of Skomer's guillemot colonies taken in the 1890s and 1930s show that the population at that time was massive, probably around 100,000 pairs. By 1945, the oceans of oil gushing from the guts of ships torpedoed during the Second World War had killed 95 per cent of southern England's Common Guillemots, Razorbills and Atlantic Puffins. Skomer's guillemot population has crept up from its low of around 2,000 pairs in the 1970s to 25,000 pairs today, and our studies show that, if nothing intervenes, fifty years from now they will have regained their 1930 levels.[26]

If nothing intervenes. Come and sit with me on the cliff top in June and look at our study colony. Scan the busy ledges with your binoculars and you'll see a mass of guillemots incubating their eggs or brooding their chicks. It all seems wonderful: guillemots at one with the world. But my longer view, based on forty-eight years of dogged year-on-year monitoring, allows me to see that Skomer's guillemots now breed some two weeks earlier than they did when I started in the 1970s. Two weeks may not sound much, but this is a huge shift. And it is not just Skomer's guillemots – many other bird species now breed much earlier than they did even twenty years ago, and the cause is climate change.[27]

At present, there seems to be no deleterious effect of this earlier breeding on Skomer's guillemots, but in other bird species climate change has generated a disastrous mismatch between when birds breed and when the food they need to rear chicks is most available. I cannot help feeling that if global warming continues it will be only a matter of time before Skomer's guillemots begin to suffer in this way. Events elsewhere show that this suffering may arrive as a sudden blast or it may come creeping year on year. Either way, continued monitoring is essential.[28]

Skomer's increasing Common Guillemot population is extraordinary because elsewhere in the north-east Atlantic numbers have crashed. The once-vast Common Guillemot populations on the Faroes, along the Norwegian coast and around Iceland have dwindled to next to nothing in the last fifty years. It is the same in Shetland, where in several recent years guillemots have simply failed to breed at all because of insufficient food. In all these areas climate change has pushed the fish on which guillemots depend out of range of their breeding colonies. It is as though you once had a supermarket just round the corner to which you walked for your week's shopping, but now the only supermarket is fifty kilometres away and you have no car. On the Isle of May in the Firth of Forth there was sufficient food for the guillemots to form eggs in 2007, but by the time those eggs hatched, the fish had gone and chicks starved to death on the ledges. A sudden surge in sea temperatures off the west coast of North America in 2015–16 delivered an unprecedented blow to the Common Guillemots there through a catastrophic loss of fish. Over 500,000 adult guillemots, and a similar number of other seabird species died.[29]

Maintaining a long-term study of guillemots, or any other species, for almost fifty years is a labour of love. There are many easier ways of passing the time. My enduring enthusiasm for guillemots is born of three things: a fascination with the way the birds have evolved to deal with social and environmental challenges that few other birds experience, their beautiful and evocative breeding places and, perhaps most of all, from a sense of responsibility to protect this species from threats of my own species' making. Often, what research biologists do seems remote from conservation. Searching for colour-rings to measure survival seems almost like an indulgence, yet survival is one of the most important measures of how well or badly a population is doing, and is especially valuable in detecting subtle, long-term changes. Identifying the species of fish that parent guillemots bring to the colony to feed their chicks is a valuable indicator of marine conditions. A few years ago, Skomer's guillemots switched from feeding their chicks on energy-rich sprats to weak and watery cod-like fish called gadids, suggesting that marine conditions had deteriorated. Luckily, after a few years, the sprats returned as the top prey item.

There are other benefits of long-term studies. The first, and a rather obvious one, is that a deep familiarity with one's study species provides a more comprehensive understanding of its biology than one would get from just a few seasons of observation. Second is the consistency with which things are done. This is important because, unless the same protocols are used each year, any year-on-year comparisons of, say, survival or chick diet are difficult to interpret. One would have thought this was common sense, but, as Peter Medawar reminded us, sense is less common than one would imagine.

I think I'd have hit it off with Alfred Newton, despite his pomposity. If he were alive today, he'd be gratified that surveys and censuses are now a routine and crucial part of ornithology; he would be delighted by the vast increase in our knowledge of birds, but he would be utterly appalled by the sheer numbers of bird species now at risk from extinction as a result of human activities. Aided by two walking sticks, he would be out there with Extinction Rebellion clamouring for a change in the way humankind operates.

Epilogue

In wildness is the preservation of the world.
Henry David Thoreau (1862)

Writing these chapters in 2021 during the Covid-19 pandemic, I am reminded of Malcolm Gladwell's *The Tipping Point*, published some twenty years ago, which likened the spread of ideas to a virus. After a lag, the virus can suddenly 'tip' and start to increase very rapidly. Some of those reviewing Gladwell's book considered it a study of 'the bleedin' obvious', but it became a bestseller nonetheless. I wonder whether the various frightening events of 2020–22 – the Covid-19 pandemic, the vast forest fires and floods caused by climate change, together with unprecedented human emigration and social unrest across the world – might trigger a tipping point in the way we treat the planet.[1]

After a succession of lucky escapes – SARS in 2003, swine flu in 2009, Ebola in 2014 – a global pandemic was almost inevitable. And after decades of denial, it suddenly seems bleedin' obvious that climate change is both real and a monumental threat. The pandemic period could also prove to be a tipping point in the way science is viewed, for without science there would have been no understanding of the Covid-19 virus and, more crucially, no vaccine with which to protect ourselves.

I hope, along with many others, that the main thing to

emerge from the pandemic will be a fundamental shift in our attitude towards nature; a greater appreciation of our reliance on a properly functioning, carefully curated planet. As is now also clear, this will depend on the outcome of a battle between the sordidly avaricious short-termists and those with a more empathetic eye to the future.

More than anything, the succession of lockdowns throughout the pandemic emphasized the central importance of social gatherings and rituals, such as Christmas, weddings and funerals, to our sense of belonging and well-being. This unwanted experiment in social deprivation on an unprecedented scale allows us to better appreciate the role of rituals in our lives and how important they must also have been to our ancestors. The images on the walls of the El Tajo cave may have been the equivalent of our festive decorations and part of a bonding ceremony whose focus was the communal consumption of a plump La Janda bustard.

The Greeks and Romans were dismissive of the Egyptians' veneration of birds. Such elitism typified the attitudes not only of colonial invaders towards indigenous peoples of other countries but also of the rich towards the poor, or among sections of national populations holding different religious beliefs. Today's archaeologists still do not understand why Palaeolithic and Neolithic peoples decorated caves and rock shelters with animal images, or what possessed the Egyptians to mummify ibises and falcons. The conquistadors made no attempt to understand the Aztec and Inca peoples' worship of birds such as the quetzal, and I have failed miserably to appreciate the bird-based stories told by the Inuit.

There's currently a move to embrace different belief systems, and even to give them equal weighting with our own scientifically orientated view of life. A more empathetic view towards cultures we do not understand is an essential antidote to the brutal colonialism that has characterized so much of human history. On the other hand, we should also ask whether all interpretations of how the natural world works are equally valid. In one sense they are, inasmuch as certain things resonate more with particular cultures, but in others, probably not. It is extremely unlikely, for example, that indigenous knowledge could have developed a prophylaxis for Covid-19, which is what science has done.[2]

We tend to think of disparities between belief systems as existing only between different countries, but they are apparent within cultures too, and often mediated by differences in wealth and educational opportunities. One example, trivial perhaps, is the disparity in understanding published bird studies by amateur versus professional ornithologists. This was brought home to me like a punch in the solar plexus when I sent a well-known television ornithologist a copy of a scientific paper my colleagues and I had written resolving the long-standing question of why Common Guillemots lay pear-shaped eggs. He told me he didn't understand a word of it. I had made the mistake of assuming he had trained as a scientist, and I was appalled by my failure to recognize how much, as a scientist, I take for granted. I immediately set about writing a more accessible account.

Indeed, a common complaint among birders is the way in which scientific studies are written and presented, and this is a good reason why scientists should also be obliged to produce so-called lay accounts of their work. The UK scientific research councils typically ask those applying for research funding to include such a lay account in their application: 'suitable for a non-specialist'. When reviewing such applications, I always read this section first, not only to better understand the aim of the study, but because it provides such a startlingly clear window into the mind of the applicant. And so often I am disappointed by the applicant's inability to express ideas simply and clearly. There is a curious paradox here. As undergraduates we know little about the ways of science, and it is only through training – a first degree and sometimes a PhD – that we eventually master the language and ways of expression that characterize the scientific approach. Once that skill has been acquired it seems almost as though it cannot be undone: the ritual of scientific training is like a mental valve allowing expression to flow in one direction only.[3]

One of the most interesting revelations I had while researching and writing this book concerned the medieval images of the Apocalypse. My largely secular upbringing meant that I had no understanding of these images other than the vague notion that they were representations of hell. As the Covid-19 death toll continued to soar through 2020 and 2021, I wondered, as many others must have done when living in the time of plague, whether we were hurtling towards the Day of Judgement. In the 1980s Barry Hines's television film *Threads* brought the horrific aftermath of a nuclear war into our living rooms and consciousness. It was particularly poignant for me since it was based in Sheffield. The breakdown in social order, the anarchy and people's desperate quest for food, so powerfully portrayed by Hines, has haunted me ever since. It is a feeling echoed by Barry Lopez:

> The desire to know ourselves better, to understand especially the source and nature of our dread, looms before us now like a spectre in a half-lit world, a weird dawn breaking

over a scene of carnage: unbreathable air, human diasporas, the Sixth Extinction, ungovernable political mobs.[4]

All of this reminded me of what the local Faroese people say in their defence of continued bird hunting. If the fishing and salmon farming on which their economy currently depends were to fail, people would be pitched into poverty, much as the Irish had been in the 1800s as a result of the potato blight. Retaining the traditional knowledge and practice of hunting seabirds (and pilot whales) is the Faroes' insurance against the future. However much you might disagree with the hunting itself, you can see their point.

If, or when, we are faced with some sort of apocalypse, our hard-won empathy for birds, little over a century old, would vanish in an instant. Empathy is a luxury born out of having sufficient resources, at least in some quarters. Wealth does not automatically result in empathy; in some cases it simply triggers the desire for more, rather like the way a fox in a hen house or a gull colony will go on killing long after it has acquired sufficient for its immediate needs. Surplus killing has evolved in foxes to allow them to make hay while the sun shines, hoarding the excess for later. I witnessed this myself one summer in Labrador when several Arctic foxes were stranded on our seabird islands, having failed to move north as the ice retreated at the end of winter. Their killing spree – puffins, mainly – just went on and on and on until the few survivors departed at the end of the season.[5] In another Labrador season, the lack of winter sea-ice (due to global warming) saw harp seals giving birth on the mainland shoreline rather than out on the sea-ice. For the local people this was manna from heaven since it allowed them to kill many more seal pups (for their pelts) than in a normal ice year.

When they told me about this unprecedented windfall I was horrified by the grasping brutality of it. Only in later years, when locals told me about their previous decades of deprivation as a result of virtual desertion by the Canadian government, did it make sense.

The role that science has played in shaping our relationships with birds is an issue that has long intrigued me. Keith Thomas, in his monumental *Man and the Natural World*, argues that the development of science in the 1600s drew us towards a more empathetic view of the natural world, away from an exploitative attitude to nature. In contrast, the novelist John Fowles in *The Tree* suggests the opposite, arguing that through their obsession with naming and describing, scientists like Linnaeus and Darwin dragged us further from nature. The low point in our relationship with nature Fowles suggests – and I agree – was during the Victorian era, as ornithologists focused on killing and collecting birds (and other animals) to stock museums in the name of science and colonialism. In his view the Victorians saw birds simply in terms of their utility – food, feathers or faeces (for fertilizer) – or as intellectual puzzles as we sought to find their place on the tree of life. The 'addiction to finding a reason, a function, a quantifiable yield, has now infiltrated all aspects of our lives – and become effectively synonymous with pleasure . . . The modern version of hell is purposelessness.'[6]

In truth, the disparity in the views of Thomas and Fowles is less than would appear at first sight. Nor are their views exclusive. As I hope I have shown, our relationship with birds throughout history has primarily been one of exploitation, but interwoven with more spiritual threads, like the ibises of ancient Egypt, the Aztecs' quetzals, the gannets on the Faroes, or the nightjars watched by Edmund Selous. I find it immensely reassuring that this more spiritual, more empathetic view of birds that started in the early years of the twentieth century is nudging utility into second place.

Although Fowles is both anti-science and anti-Darwin, my feeling is that science, especially the fields of animal behaviour, ecology and evolution, all of which started with Darwin, has brought us much closer to birds. Fowles's view, with which I have much sympathy, is myopic inasmuch that it overlooks both the millennia of avian exploitation prior to the Victorian era and the change in attitude that started around 1900. By focusing on the Victorians, Fowles – whose name literally means 'wild birds' – creates a straw man.

Science is a curious process, often generating in the popular imagination a dispassionate, unfeeling beast. By necessity, science forces us to be objective and critical in the appraisal of evidence, but this hides the fact that many scientists who study birds started out as birdwatchers and have retained – and in many cases developed – a marked empathy for birds such that they revel in the spiritual experience of being among them and studying their habits. In addition, some of the so-called 'new nature writing' has successfully merged objective and subjective responses to nature, creating a bridge between C. P. Snow's 'two cultures'.[7]

It seems a remarkable coincidence that the term 'empathy' entered the English language exactly at the time – the early 1900s – that people like Florence Merriam Bailey in the US and Edmund Selous in Britain were championing more empathetic attitudes towards birds. Psychologists recognize a continuum, from cold 'cognitive empathy' at one extreme to tearful 'emotional empathy' at the other, with 'compassionate empathy' – a balance between logic and emotion – in between. Like many other personality traits, both genes and upbringing determine empathy. Research shows that compassionate empathy can be learned, but also that people find it much easier to behave empathetically towards those with similar beliefs and values, making Bernardino de Sahagún's and Francisco Hernández's rapport with the indigenous people of Mexico (Chapter 6) especially remarkable.

We live in an age where empathy for birds and the natural world has never been greater, but if the Covid-19 pandemic is to precipitate a tipping point in our concern for the planet, we need even more of that kind of empathy. It is not going to be easy. And it is especially ironic that our increasing awareness of the psychological benefits of human–nature interactions coincides so spectacularly with a reduction in the opportunities for access to the natural world.[8] Of the several ways we might rectify this, one is through increasing knowledge. It is well known that a visit to an art gallery or an archaeological site is so much more rewarding if we know something about those who created what is on view. In a similar way, knowing about the behaviour, ecology, structure and evolution of birds can have a transformative effect on the way we value them. I have felt some kind of empathy towards birds ever since I was young, but an incident in my early twenties during my PhD studies on Skomer elevated my empathy to a new level.

This epiphanic experience occurred one day when I realized that the Common Guillemot I was watching on the ledge in front of me had recognized its partner several hundred metres away out at sea, when to me it seemed no more than a brown speck. As the bird flew in and arrived beside its mate, I knew that, despite all my experience, I had underestimated their abilities. This single incident forced me to imagine what it was like to be a bird. Not everyone is likely to have that kind of experience, but talking about it, writing about it and teaching undergraduates about it can help to inspire others.

The recent rise of young environmental activists is encouraging, and it is they who are most likely to trigger the tipping point over concern for the planet. Their motivation is a mixture of fear – of what happens if we do nothing – and wonder – at how the natural world works – but most importantly, they appreciate the spectacle that is the natural world. Wonder is also the way the separate strands of science and empathy can be woven together.

One of my most rewarding relationships with birds occurred in 2018, when I introduced a group of undergraduates to birds for the first time. I had been awarded a teaching prize that came with a stipend to be spent on some aspect of teaching. At the awards ceremony in London each of us was asked to say how we would spend the money. As, one after the other, the awardees described ways they would modify or reinvigorate their classroom teaching, I was worried that my plan to take my entire first-year class to see the seabird colony at Bempton would be met with disapproval. But it wasn't.

The day came. The coaches lined up outside the university library: 8 a.m. is early for undergraduates, but they were all there. Two hours later we were at the colony and walking along the top of the 400ft gleaming chalk cliffs thronged with birds, under an azure sky: kittiwakes, guillemots, puffins, fulmars, gannets . . . the sight and smell and sound of the birds were all-consuming, and I could sense the students' excitement. Previously, I had briefed them about the unusual biology and life history of seabirds and how, globally, their number had halved in the last fifty years. I had also arranged for a colleague, Keith Clarkson, to join us. I had taught Keith as an undergraduate in the 1970s – it was he who refused to disclose the location of the local Peregrines (Chapter 10). He was now site manager at the Bempton Reserve on the Yorkshire coast. Keith is an utterly charismatic enthusiast, and I glowed with pride and excitement as I saw him lighting fires in the students' minds.

A short additional coach ride took us to Bridlington harbour, where we boarded the pleasure cruiser *Yorkshire Belle* to view the seabirds from the sea. As we gently nosed in under the cliffs, the boat's skipper provided us with unparalleled proximity to a mass of birds that seemed remarkably unperturbed by our presence. It felt as though we were now part of that throbbing super-organism that is a seabird colony: vibrant, cacophonous, smelly, pulsating with extraordinary life. Flocks of Common Guillemots each carrying a single sprat lengthways in their bill streamed past; parties of Atlantic Puffins and Razorbills swam beneath the bows of the boat; and in the air above us, Northern Gannets soared like pterodactyls. To see the expressions on the hundred undergraduate faces was among the most gratifying experiences of my forty years of teaching, and I will be more than satisfied if one or two took away with them a greater empathy for birds.

List of Birds Mentioned in the Text

These are the common and scientific names of birds from the IOC World Bird List v11.2: https://www.worldbirdnames.org/new/ioc-lists/master-list-2/). The list here is in alphabetical order.

Acorn Woodpecker *Melanerpes formicivorus*
African Blue Quail *Synoicus adansonii*
African Finfoot *Podica senegalensis*
American Flamingo *Phoenicopterus ruber*
American Kestrel *Falco sparverius*
Arabian Babbler *Argya squamiceps*
Atlantic Puffin *Fratercula arctica*
Avocet – see Pied Avocet
Bald Eagle *Haliaeetus leucocephalus*
Bar-tailed Godwit *Limosa lapponica*
Barbary Partridge *Alectoris barbara*
Barn Owl – see Western Barn Owl
Barn Swallow *Hirundo rustica*
Bearded Vulture *Gypaetus barbatus*
Bengalese Finch – see White-rumped Munia
Bittern – see Eurasian Bittern
Black-crowned Night Heron *Nycticorax nycticorax*
Black-faced Honeycreeper *Melamprosops phaeosoma*
Black Francolin *Francolinus francolinus*
Black Grouse *Lyurus texrix*
Black Guillemot *Cepphus grylle*
Black Kite *Milvus migrans*
Black-legged Kittiwake *Rissa tridactyla*
Blackbird – see Common Blackbird
Blackcap – see Eurasian Blackcap
Blue Rock Thrush *Monticola solitarus*

Blue-tailed Hummingbird – see Long-tailed Sylph
Blue-throated Hillstar *Oreotrochilus cyanolaemus*
Blue Tit – see Eurasian Blue Tit
Boat-billed Flycatcher *Megarynchus pitangua*
Bonxie – see Great Skua
Brambling *Fringilla montifringilla*
Brent Goose/Brant Goose *Branta bernica*
Brünnich's Guillemot/Thick-billed Murre *Uria lomvia*
Bullfinch – see Eurasian Bullfinch
Bulo Burti Shrike (or Boubou Shrike) – see Coastal Boubou
Canary – see Domestic Canary
Carmine Bee-eater – see Southern Carmine Bee-eater
Carrion Crow *Corvus corone*
cassowary *Casuarius* spp.
Cattle Egret *Bubulcus ibis*
Chiffchaff – see Common Chiffchaff
Chough – see Red-billed Chough
Coastal Boubou *Laniarius nigerrimus*
Comb-crested Jacana *Irediparra gallinacea*
Common Blackbird *Turdus merula*
Common Buzzard *Buteo buteo*
Common Chiffchaff *Phylloscopus collybita*
Common Crane *Grus grus*
Common Cuckoo *Cuculus canorus*
Common Eider *Somateria mollissima*
Common Guillemot/Murre *Uria aalge*
Common Gull – see Mew Gull
Common Kestrel *Falco tinnunculus*
Common Kingfisher *Alcedo atthis*
Common Loon – see Great Northern Diver
Common Moorhen *Gallinula chloropus*
Common Nightingale *Luscinia megarhynchos*
Common Ostrich *Struthio camelus*
Common Pheasant *Phasianus colchicus*
Common Quail *Coturnix coturnix*
Common Redshank *Tringa tetanus*

Common Redstart *Phoenicurus phoenicurus*
Common Rock Thrush *Monticola saxatalis*
Common Scoter *Melanitta nigra*
Common Shelduck *Tadorna tadorna*
Common Snipe *Gallinago gallinago*
Common Starling *Sturnus vulgaris*
Common Woodpigeon *Columba palumbus*
Coot – see Eurasian Coot
Cormorant – see Great Cormorant
Corncrake *Crex crex*
Crested Coquette Hummingbird – see Rufous-crested Coquette
Crested Parrot – see Raven Parrot
Cuckoo – see Common Cuckoo
Cuckoo-finch *Anomalospiza imberbis*
Curlew – see Eurasian Curlew
Dalmatian Pelican *Pelecanus crispus*
Darwin's Rhea *Rhea darwinii*
Demoiselle Crane *Grus virgo*
Dickcissel *Spiza americana*
Dodo *Raphus cucullatus*
Domestic Canary *Serinus canaria*
Domestic Fowl *Gallus domesticus*
Domestic Pigeon – see Feral Pigeon
Dotterel – see Eurasian Dotterel
Dunnock *Prunella modularis*
Eider – see Common Eider
Eurasian Bittern *Botarus stellaris*
Eurasian Blackcap *Sylvia atricapilla*
Eurasian Blue Tit *Cysanistes caeruleus*
Eurasian Bullfinch *Pyrrhula pyrrhula*
Eurasian Coot *Fulica atra*
Eurasian Curlew *Numenius arquata*
Eurasian Dotterel *Charadrius morinellus*
Eurasian Golden Oriole *Oriolus oriolus*
Eurasian Hoopoe *Upupa epops*
Eurasian Jay *Garrulus glandarius*
Eurasian Magpie *Pica pica*
Eurasian Oystercatcher *Haematopus ostralegus*
Eurasian Siskin *Spinus spinus*
Eurasian Skylark *Alauda arvensis*
Eurasian Sparrowhawk *Accipiter nisus*

Eurasian Spoonbill *Platalea leucorodia*
Eurasian Teal *Anas crecca*
Eurasian Tree Sparrow *Passer montanus*
Eurasian Whimbrel *Nemenius phaeopus*
Eurasian Wigeon *Mareca penelope*
Eurasian Woodcock *Scolopax rusticola*
Eurasian Wryneck *Jynx torquilla*
European Bee-eater *Merops apiaster*
European Golden Plover *Pluvialis apricaria*
European Goldfinch *Carduelis carduelis*
European Green Woodpecker *Picus viridis*
European Herring Gull *Larus argentatus*
European Honey Buzzard *Pernis apivorus*
European Nightjar *Caprimulgus europaeus*
European Robin *Erithacus rubecula*
European Shag *Phalacrocorax aristotelis*
European Stonechat *Saxicola rubicola*
European Storm Petrel *Hydrobates pelagicus*
European Turtle Dove *Streptopelia turtur*
fairywren – *Malurus* spp.
Feral Pigeon *Columba livia domestica*
Fieldfare *Turdus pilaris*
Flamingo – see Greater and American Flamingo
Fox Sparrow *Passerella* spp.
Fulmar – see Northern Fulmar
Gannet – see Northern Gannet
Giant Kingfisher *Megaceryle maxima*
Glossy Ibis *Plegadis falcinellus*
Goldcrest *Regulus regulus*
Golden Eagle *Aquila chrysaetos*
Golden Oriole – see Eurasian Golden Oriole
Golden Plover – see European Golden Plover
Goldfinch – see European Goldfinch
Goshawk – see Northern Goshawk
Great Auk *Pinguinus impennis*
Great Bustard *Otis tarda*
Great Cormorant *Phalacrocorax carbo*

Great Crested Grebe *Podiceps cristatus*
Great Curassow *Crax rubra*
Great Northern Diver *Gavia immer*
Great Skua *Stercorarius skua*
Great White Pelican *Pelecanus onocrotalus*
Great Spotted Woodpecker *Dendrocopus major*
Greater Flamingo *Phoenicopterus roseus*
Greater Honeyguide *Indicator indicator*
Greater Painted Snipe *Rostratula benghalensis*
Green Woodpecker –see European Green Woodpecker
Grey Heron *Ardea cinerea*
Grey Partridge *Perdix perdix*
Griffon Vulture *Gyps fulvus*
Guadalupe Caracara *Caracara lutosa*
Guillemot – see Common Guillemot/Murre
Guira Cuckoo *Guira guira*
Gurney's Pitta *Hydronis gurneyi*
Gyrfalcon *Falco rusticola*
Harpy Eagle *Harpia harpyja*
Herring Gull – see European Herring Gull
Hobby *Falco subbuteo*
Honey Buzzard – see European Honey buzzard
Hoopoe – see Eurasian Hoopoe
House Sparrow *Passer domesticus*
Hume's Leaf Warbler *Phylloscopus humei*
Hume's Short-toed Lark *Calandrella acutirostris*
Hume's Treecreeper *Certhia manipurensis*
Hume's Wheatear *Oenanthe albonigra*
Indian Peafowl *Pavo cristatus*
Ivory-billed Woodpecker *Campephilus principalis*
Jack Snipe *Lymnocryptes minimus*
Jackdaw – see Western Jackdaw
Jay – see Eurasian Jay
Junglefowl – see Red Junglefowl
Kestrel – see Common Kestrel
King Bird-of-Paradise *Cicinnurus regius*
Kingfisher – see Common Kingfisher
Kite – see Black Kite
Kittiwake – see Black-legged Kittiwake
Kori Bustard *Ardeotis kori*

Lammergeier – see Bearded Vulture
Lapwing – see Northern Lapwing
Lesser Rhea *Rhea pennata*
Lesser Spotted Woodpecker *Dryobates minor*
Little Bee-eater *Merops pusillus*
Little Bustard *Tetrax tetrax*
Long-tailed Paradise Whydah *Vidua paradisea*
Long-tailed Sylph *Aglaiocercus kingii*
Long-tailed Tit *Aegithalos caudatus*
Long-tailed Widowbird *Euplectes progne*
Lovely Cotinga *Cotinga amabilis*
Magpie – see Eurasian Magpie
Malachite Sunbird *Nectarina famosa*
Manx Shearwater *Puffinus puffinus*
Marvellous Spatuletail *Loddigesia mirabilis*
Meadow Pipit *Anthus pratensis*
meadowlark *Sturnella* spp.
Merlin *Falco columbarius*
Mew Gull *Larus canus*
Mexican House Finch (aka House Finch) *Carpodacus mexicanus*
Middle Spotted Woodpecker *Dendrocoptes medius*
Moorhen – see Common Moorhen
Mute Swan *Cygnus olor*
Nightingale – see Common Nightingale
Nightjar – see European Nightjar
Northern Bald Ibis *Geronticus eremita*
Northern Cardinal *Cardinalis cardinalis*
Northern Fulmar *Fulmarus glacialis*
Northern Gannet *Morus bassanus*
Northern Goshawk *Accipiter gentilis*
Northern Lapwing *Vanellus vanellus*
Northern Pintail *Anas acuta*
Northern Raven *Corvus corax*
Orange Fruit Dove *Ptilinopus victor*
Ortolan Bunting *Emberiza hortulana*
Ostrich – see Common Ostrich
Painted Snipe – see Greater Painted Snipe
Paradise Whydah – see Long-tailed Paradise Whydah

partridge *Alectoris* spp.
Peacock – see Indian Peafowl
Peregrine Falcon *Falco peregrinus*
Pheasant – see Common Pheasant
Pied Avocet *Recurvirostra avosetta*
Pied Kingfisher *Ceryle rudis*
Pin-tailed Sandgrouse *Pterocles alchata*
Pin-tailed Whydah *Vidua macroura*
Pintail – see Northern Pintail
Puffin – see Atlantic Puffin
Purple Gallinule – see Western Swamphen
Quail – see Common Quail
Quetzal – see Resplendent Quetzal
Raven – see Northern Raven
Raven Parrot *Lophopsittacus mauritianus*
Razorbill *Alca torda*
Red-backed Shrike *Lanius collurio*
Red-billed Chough *Pyrrhocorax pyrrhocorax*
Red Junglefowl *Gallus gallus*
Red Knot *Calidris canutus*
Red-legged Partridge *Alectoris rufa*
Resplendent Quetzal *Pharomachrus mocinno*
Rhea – see Darwin's Rhea/ Lesser Rhea
Robin – see European Robin
Rodrigues Solitaire *Pezophaps solitaria*
Rook *Corvus frugilegus*
Ruby-topaz Hummingbird *Chrysolampis moquitus*
Ruddy Shelduck *Tadorna ferruginea*
Ruff *Calidris pugnax*
Rufous-crested Coquette *Lophornis delattrei*
Sacred Ibis *Threskiornis aethiopicus*
Scarlet Ibis *Eudocimus ruber*
Siberian Crane *Leucogeranus leucogeranus*
Siskin – see Eurasian Siskin
Skylark – see Eurasian Skylark
Snethlage's Tody-tyrant *Hemitriccus minor*
Snipe – see Common Snipe
Solitaire – see Rodrigues Solitaire
Song Sparrow *Melospiza melodia*
Southern Carmine Bee-eater *Merops nubicoides*
Sparrowhawk – see Eurasian Sparrowhawk

Spatule-tailed Hummingbird – see Marvellous Spatuletail
Spoonbill – see Eurasian Spoonbill
Spoon-billed Sandpiper *Calidris pygmaea*
Spotted Sandpiper *Actitis macularius*
Squirrel Cuckoo *Piaya cayana*
Starling – see Common Starling
Stonechat – see European Stonechat
Storm Petrel – see European Storm Petrel
Tawny-flanked Prinia *Prinia subflava*
Tawny Owl *Strix aluco*
Teal – see Eurasian Teal
Toco Toucan *Rhamphastos toco*
Tree Sparrow – see Eurasian Tree Sparrow
Turkey – see Wild Turkey
Turtle Dove – see European Turtle Dove
Virginian Nightingale – see Northern Cardinal
Western Barn Owl *Tyto alba*
Western Jackdaw *Corvus monedula*
Western Osprey *Pandion haliaetus*
Western Swamphen *Porphyrio porphyrio*
Whimbrel – see Eurasian Whimbrel
White-faced Epthianura – see White-fronted Chat
White-fronted Chat *Epthianura albifrons*
White Pelican – see Great White Pelican
White-rumped Munia *Lonchura striata*
White Stork *Ciconia ciconia*
White-throated Dipper *Cinclus cinclus*
White's Thrush *Zoothera aurea*
widowbird *Euplectes* spp.
Wigeon – see Eurasian Wigeon
Wild Turkey *Meleagris gallopavo*
Willow Warbler *Phylloscopus trochilus*
Wood Pigeon – see Common Woodpigeon
Wood Warbler *Phylloscopus sibilatrix*
Woodcock – see Eurasian Woodcock
Wren *Troglodytes troglodytes*
Wryneck – see Eurasian Wryneck
Zebra Finch *Taeniopygia guttata*

Acknowledgements

The last person to know everything lived a long time ago, and since then what we know about the world of birds and how they intensify our lives has continued to grow. We are lucky to live in a world rich in experts, and I have been fortunate to be able to seek their advice on some of the topics I discuss here in *Birds and Us*. Throughout the research and writing I have been blessed with good advice and this is my opportunity to thank the many people that have helped me.

My first thanks – heartfelt thanks – are to my agent Felicity Bryan for her gentle, but relentless encouragement to embark on this project. I'm only sorry she did not live long enough to see the fruits of her efforts.

To avoid this becoming an amorphous sea of names, I've arranged them mainly by chapter, starting on a gloriously sunny morning in southern Spain in 2017 when I met up with María Lazarich González, Antonio Ramos-Gil and Carmen Fernandez from the University of Cadiz, to visit the El Tajo de las Figuras rock shelter. I thank the Junta de Andalusia for granting permission for the visit. Maria's enthusiasm and generosity in sharing her knowledge and images has been truly inspirational. Others who helped in various ways with this chapter include Francisco Valera Hernández, Geoff Hancock, José Manuel López Vázquez (Presidente Asociación de Amigos de la Laguna de la Janda), Martí Mas Cornellà, Jean Clottes, Jim Whitaker (who published Willoughby Verner's diaries), Andrea Pilastro and Paul Pettitt, Professor of Anthropology at the University of Durham in the UK, who together with Penny Wilson generously hosted a one-day meeting on birds in cave art for my benefit.

For Egyptian expertise I am grateful to R. B. Parkinson, Paul Nicholson, Nigel-Harcourt-Brown, Rozenn Bailleul-LeSuer, Douglas Russell and Joanne Cooper at the Natural History Museum in Tring, who allowed me to examine ibis and hawk mummies, and to Lidija McKnight at the University of Manchester for telling me about her bird mummy research.

My friend Jeremy Mynott provided much helpful advice on the Greeks and Romans. I am especially grateful to the various people that helped to make it possible for me to examine the tongue of a flamingo, most notably Juan Amat, but also Nicola Hemmings, Herman Berkhoudt, Rick Wright, Arnaud Bechet, and Nicola Baccetti, whose offer for us to share some flamingo tongues I still hope to take up.

Medieval falconry informants to whom I am grateful include Baudouin Van den Abeele, Matt Gage, Karl-Heinz Germann and Michael Warren. Thanks also to those who helped me track down details of W. B. Yapp: Jim Reynolds, Lawrie Finlayson, Pat Butler, Jeremy Greenwood, Nick Davies, Rhys Green, Chris Perrins, Ian Newton, Carole Showell (Librarian at the British Trust for Ornithology), Jane Whittle and Yapp's niece, Katharine Butterworth.

I am grateful to my colleagues who comprised the Willughby International Network, funded by the Leverhulme Trust between 2012 and 2016, which gave me an opportunity to better understand Francis Willughby, John Ray and their world, in particular, Dorothy Johnston, Mark Greengrass, Isabelle Charmantier, Sachiko Kusukawa, Paul J. Smith, Ann-Marie Roos and Carlo Violani. For information and advice regarding the New World I thank Alexander Lees, Amy Buono, Elke Bujok, Christine Jackson, Marcy Norton, Penny

Olsen, Sebestian Kroup, Paul Haemig, Chris Preston, Rebecca Tillett and Ulina Rublack.

In the Faroes I was generously hosted by Sjurdur Hammer and his family and Inga Tórarenni, who together with Jens-Kjeld Jensen, Bergur Olsen, Hans Andrias Sølvará, Jákup Reinert Hansen, Kim Simonsen, Marita Gulklett, Jamie Thompson, Edward Fuglø and the residents of Skúvoy (who allowed me to witness their chain dance), made this an extraordinarily exciting trip. For other seabird information relating to this chapter I am extremely grateful to Sarah Wanless, Mike Harris, John Love and Tycho Anker-Nilssen.

I thank John Holmes for an all-too brief conversation following his stimulating talk on the Pre-Raphaelites' links with science, while he was in Sheffield.

Robert Prys-Jones (whom I first met as a PhD student in the early 1970s), Julian Hume, Nigel Collar and Shyamal Lakshminarayanan provided valuable advice relating to Allan O. Hume. Jonathan Elphick and Christine Jackson provided helpful information about bird artists. Thanks also to Douglas Russell and numerous other museum curators for sharing their views on the essential value of museum collections.

Many of my birding colleagues, both birdwatchers and bird-watchers, provided me with information, notably Bryan Barnacle, Leo Batten, Ken Blake, Neil Bucknell, Andy Clarke, Mark Cocker, John Eyre, Roger Jolliffe, Dick Newell, Richard Porter, Roger Riddington and David Wood.

It is a pleasure to thank several people for sharing their knowledge of the history of Bempton and its seabirds: Elizabeth Boardman, Linda Ellis, Margaret Traves and especially Keith Clarkson.

Then there are those people whose help extended beyond particular chapters, to whom I am especially grateful: Sue

Barnes, Patricia Brekke, Chris Everest (who before his retirement from Sheffield University Library, tracked down many obscure references for me – as did his successors), Nicola Hemmings, Alan Knox, Scott Pitnick, Tom Pizzari and Ann Sylph (Zoological Society of London). Special thanks are due to four friends for their help, stimulation and support: Bill Swainson, Karl Schulze-Hagen, Bob Montgomerie and Jeremy Mynott. Bob and Jeremy both read the entire manuscript and I thank them for their critical comments as well as for years of wise counsel. I am especially grateful to Trevor Horwood for his excellent copy-editing, to Kit Shepherd for assiduous proofreading and to my beta readers for their comments. Thanks too for enthusiastic support from Caspar Dullaart at Atlas Contact Publishers, my agent Carrie Plitt, and my wonderful editor Tom Killingbeck and his team at Viking/Penguin.

Most of my previous books were written, at least in part, among the birds and mountains of southern Spain; this one, while still inspired by that region, was written largely at home in Sheffield under the restrictions imposed by Covid and I thank my wife Miriam for her fortitude during this time.

Notes

Chapter 1: Of Peculiar Interest: Neolithic Birds

1 Verner (1914).
2 See Broderick (1963) for a biography of Breuil. Bahn (2016: 67). The Pileta cave and its drawings were discovered by José Bullón Lobato in 1905: the Bullón family still own the cave and run the popular tours – 12,000 visitors a year – the youngest Bullón told me in May 2019. Verner's account of his visit to the Pileta cave is in the *Saturday Review* 1911 in six parts (Verner, 1911). Verner's account of the El Tajo cave is in *Country Life* 1914 in three parts (Verner, 1914).
3 Breuil and Burkitt (1929). Breuil and his protégé, Cambridge archaeologist Miles Burkitt, visited El Tajo in 1914 (Díaz-Andreu, 2013), later producing their book, with Breuil providing the main text, Burkitt writing the introduction and conclusions and Sir Montagu Pollock undertaking 'the arduous task of translating, and in many cases indeed of paraphrasing, the condensed descriptions of our French colleague'. Willoughby Verner should also have been a co-author but was not included because of an unspecified 'misunderstanding' (Perelló, 1988: 179; Martí Mas Cornellá, pers. comm., and Cornellá, 2003–5). There was also a brief account of El Tajo by Molina (1913); see also Lazarich et al. (2019).
4 Chapman and Buck (1910: 242); Macpherson (1896).
5 de Juana and Garcia (2015) and E. Garcia pers. comm.
6 Clottes (2016).
7 Ibid.

Chapter 2: Inside the Catacombs: The Birds of Ancient Egypt

1 Bailleul-LeSuer (2013: 16).
2 Markham (1621) and Chapman (1930).
3 McCouat (2015); the quote is from Parkinson (2008: 9).
4 McCouat (2015).
5 Moser (2020).
6 Ibid.
7 Vansleb (1678).
8 Wilde (1840).
9 Browne's 'Fragment on mummies' in Wilkin (1835: 276).
10 Smith (2018).
11 Usick (2007); Manley and Ree (2001: 154).
12 According to Martin (1981: 4).
13 Houlihan (1986: 140); von den Driesch et al. (2005).
14 Herodotus cited in Mynott (2018: 200–201).
15 Landauer (1961); see also Ray (1678); Ikram (2015: 12).
16 Réaumur (1750).
17 Cuvier (1817).
18 Wasef et al. (2019).

Chapter 3: Talking Birds: The Beginnings of Science in Greece and Rome

1 Smellie (1790).
2 Aristotle (1943); Leroi (2014); Birkhead and Lessells (1988); Birkhead et al. (1988). The high frequency of copulation in raptors may be a way male raptors help to protect their paternity; copulation may also be part of pair bonding. Some birds – the Common Blackbird, for example – copulate only a handful of times for each clutch.

3 Pizzari et al. (2003).
4 Brock (2004).
5 Birkhead et al. (1995). There's an important point here: last male sperm precedence in birds occurs only when two males inseminate a female one after the other, as in Aristotle's replacement observation. With wild birds this would be relatively rare, and extra-pair copulations would typically occur at any time in the copulation period. Their success may be enhanced by the transfer of more or better sperm. See also Brock (2004).
6 Leroi (2014).
7 Mynott (2018: 224).
8 Medawar and Medawar (1983); Charles Darwin's letter of 22 February 1882 to William Ogle; Leroi (2014).
9 Newmyer (2011).
10 Aristotle (1965: 536b); Mynott (2018: 142).
11 Plutarch on the intelligence of animals, cited in Kleczkowska (2015: 99).
12 West and King (1990).
13 In the opening lines of thc third book of the *Iliad*, Homer writes that the cries of the Trojans resembled the clangour of thc cranes, fleeing across the heavens.
14 Cited in Mynott (2018: 65).
15 Genesis 1: 26.
16 Villing et al. (2013).
17 Singer (1931).
18 Lack (1968).
19 Davies (2015).
20 Ibid.
21 Dakin and Montgomerie (2013).
22 Pliny (1885).
23 Mynott (2009).

24 The flamingo's unusual head, with its large 'Roman' nose, has evolved to allow them to exploit an unusual ecological niche: tiny food items in water. In most birds the upper mandible of the beak is larger than the lower, but in the six species of flamingo it is the other way round, in part because the head is held upside down in the water during feeding. Associated with this, it is the flamingo's upper mandible that moves while the lower one remains stationary – the opposite of most other birds. As the upper mandible opens and shuts, the tongue moves forward and back, pumping water through the mouth, allowing shrimps and diatoms to be filtered out by the sieve-like lamellae and then swallowed.

25 Buffon (1770–83, vol. VIII: 446) trans. W. Smellie, 1792–3.

26 Ibid.; see also Mynott (2018: 106–7).

27 Chapman and Buck (1910).

28 N. Baccetti, pers. comm. The bird I dissected was also a power-lines victim.

29 Mynott (2018).

30 Ibid.

31 Pliny (1885).

Chapter 4: Manly Pursuits: Hunting and Conspicuous Consumption

1 Robinson (2003).

2 Other academics to have studied medieval birds include the British biologist Evelyn Hutchinson FRS, sometimes referred to as the father of modern ecology (Hutchinson, 1974), and the American, Charles Vaurie, best known for his work on avian systematics (Vaurie, 1971).

3 Yapp (1987); Cummins (1988). The 'kestrel for a knave' story had its origin in the first printed book on falconry – *The Boke of St Albans* – by Juliana Berners in 1486, and is widely cited as the source of the idea that a hierarchy existed between the raptors and their owners: 'An Eagle for an Emperor, a Gyrfalcon for a King, a Peregrine for a Prince, a Saker for a Knight, a Merlin for a Lady, a Goshawk for a Yeoman, a Sparrowhawk for a Priest, a Musket for a Holy-water Clerk, a Kestrel for a Knave.' However, the 'kestrel for a knave' appears only in another fifteenth-century MS, and, as several successive writers have pointed out, the hierarchy of birds and people was unlikely to exist in reality (Oggins 2004).

4 Yapp (1981).

5 Yapp (1987); Owen-Crocker (2005); Bloch (2005).

6 Cade (1982); Newton and Olsen (1990); Robinson (2003: 14); Mynott (2018: 154); Aristotle (1936: 27.118).

7 Canby (2002: 166).

8 This Portuguese mosaic, known as the 'Mosaico de Cavaleiro', is in the south-eastern town of Mértola: see Macias (2011).

9 This caution was aimed especially at those attending sick hawks (Robinson 2003: 28).

10 Canby (2002: 166); and Macdonald's comment on Cade: www.ef-fc.org/projects, February 2019; Oggins (2004: 35).

11 Oggins (2004: 35).

12 Fox (1995).

13 Strutt (1842: 24).

14 Wood and Fyfe (1943); and see Oggins (2004: 127).

15 James (1925).

16 Yapp (1979).

17 To me the birds look like the African Blue Quail, but this seems unlikely: see Endpapers.

18 Bibles were so heavy they required a lectern (meaning 'to read') for support.

19 See Hutchinson (1974).

20 Yapp (1981: 75).

21 Briggs (2014).

22 Revelation 19: 17–18.

23 Wood and Fyfe (1943).

24 Ibid.

25 Willemsen (1943); Haskins (1921); Vaurie (1971); Venturi (1904); Yapp (1983). It has been suggested that at least some of the birds were copied from the sixth-century illuminated manuscript known as the 'Vienna Dioscorides' (see p. 84–5), but once again Yapp (1983), with forensic thoroughness, compared the two and found no link.

26 Venturi (1904), Yapp (1983). The entire Vatican MS is reproduced online at http://digi.vatlib.it/view/MSS-Pal.lat.1071.

27 Allen (1951: 398).

28 Haffer et al. (2014).

29 Oggins (2004: 129).

30 Ibid.

31 Oggins (2004: citing various authors in note 133).

32 The hood was introduced in the thirteenth century (Robinson 2003).

33 This is from Frederick II's book: see Wood and Fyfe (1943).

34 Oggins (2004: 32).

35 The same method was used to catch Rooks: 'Take some thick brown paper, and divide a sheet into eight parts, and make them up like sugar-loaves; then lime the inside of the paper . . . then put some corn in them, and lay three score or more of them up and down the ground . . . then stand at a distance, and you will see most excellent sport; for as soon as rooks . . . come to peck out any of the corn, it will hang upon his head,

and he will immediately fly bolt upright so high that he should soar almost out of sight; and when he is spent, come tumbling down, as if they had been shot' (Cox 1686).

36 Shaw (*Speculum Mundi*, 1635, cited in Harting (1871: 223)).

37 Freeman and Salvin (1859), cited in Harting (1871: 24).

38 The species of falcon isn't specified: Freeman and Salvin (1859): https://archive.org/details/FalconryItsClaims/page/n165/mode/2up/search/Sultan

39 'Are minds . . .' cited in Bates (2011: 404; see also 420). Margaret Cavendish wrote an anti-cruelty poem entitled 'The Hunting of the Hare' in 1653: http://library2.utm.utoronto.ca/poemsand-fancies/2019/04/29/the-hunting-of-the-hare/. Gascoigne's book is usually bound with Turberville's *Booke of Faulconrie* (1575), and for some time it was thought that Turberville was the author (or compiler) of both (Bates 2011: 403). Anti-hunting sentiments are discussed in Keith Thomas (1983) and others.

40 Harwood (1928). An obvious exception, of course, was pets (see Thomas 1983: 100).

41 Lecky (1913).

42 Thomas (1983).

43 Strutt (1842: 25). Charles II was notoriously promiscuous: 'Restless, he rolls about from whore to whore' – J. Wilmot in Miller (1991).

44 Jacob (1718); see also Strutt (1842: 25); Nash (1633) on the frivolity of hunting – 'an unnatural pastime'; Nash is disturbed by falconry's apparent 'approval' of the fact that female raptors are larger and more powerful than their male partners.

45 See https://iaf.org/ethical-and-scientific-aspects-concerning-animal-welfare-and-falconry/.

46 Katherine Butterworth (W. B. Yapp's niece), pers. comm., quoted with permission.

Chapter 5: Renaissance Thinking: The Parts of Birds

1 Italian poet Alessandra Scala asks this of her correspondent, Italy's most renowned female fifteenth-century scholar, Cassandra Fedele, in 1491, characterizing what is clearly an age-old problem for women: the choice between a career or a family. The quote is cited in King (1980). Also M. L. King, pers. comm.

2 Isaacson (2017: 178 and 398).

3 Painting attributed to Caravaggio, see Jackson (1993); eating woodpeckers, see Muffett (1655) and Naumann and Naumann (1820–60). Aristotle commented that the Wryneck's tongue was equal to four fingers' breadth in length, but makes no mention of the woodpecker's tongue (cited in Lones 1912: 181, 247).

4 Bock (2015). Darwin in *On the Origin of Species* used woodpeckers as an example to illustrate the origin of adaptations.

5 Belon (1555).

6 Lones (1912).

7 Ghosh (2015).

8 Ibid.

9 Isaacson (2017: 423).

10 Cole (1944: 56); Richardson (1885).

11 Allen (1951: 410). Coiter volunteered to served as an army physician in the French Wars of Religion, and died of an (unknown) illness in 1576 less than a month after the Edict Beaulieu, which ended the Fifth War of Religion.

12 Oviedo (1526).

13 Belon's Swiss contemporary Conrad Gessner, saw a toucan only after he had completed his own encyclopedia the same year, 1555. Gessner subsequently added the toucan to his later publication *Icones avium omnium* in 1560. The Toco Toucan does have nostrils, but they are hidden.

14 The arrangement of toes was later used in avian classification. However, recent research has shown that zygodactyly (*zygo* in Greek meaning 'even') has evolved independently no fewer than nine times in birds, so it must be used in classification with care. The most common arrangement of toes is three toes forward, one back (anisodactyly).

15 Smith (2007).

16 Lovegrove (2007).

17 Ibid.: 29.

18 In Britain and central Europe, the Rook has long been regarded as a pest of newly sown spring cereals. In Scotland, where Rooks were especially abundant, James II introduced legislation in 1457 designed to prevent Rooks from breeding and to reduce their numbers. They were killed by poisoning, shooting and the destruction of nests. In some areas the shooting of colonies became a kind of folk festival, with music and dancing. The only way to keep Rooks off recently sown cereals was to employ people – children, the disabled and the aged – to scare the birds away. By the 1750s it was realized that, by consuming a wide range of soil invertebrates, Rooks probably did more good than harm. Even so, by the beginning of the twentieth century plans were afoot to eliminate the Rook altogether from certain areas of Germany. The plan almost succeeded, with numbers in northern Germany falling from 35,000 pairs to just a few hundred by the 1950s. After the killing ceased in the 1970s, numbers recovered (Krüger et al., 2020).

19 The sparrow-killing campaign stopped in 1960 as a result of the Chinese ornithologist Tso-hsin Cheng pointing out to the government that the birds ate harmful insects as well as grain. The Chinese government subsequently imported a quarter of a million sparrows from the Soviet Union in an

effort to restore the ecological balance (Pantsov and Levine, 2013).

20 The accounts were examined by a Norfolk ornithologist Daniel Gurney Sr in 1834. The spelling of Lestrange varies from le Straunge and Le Strange to L'Strange.

21 Gurney (1921: 132). Newton (1896: 738) suggests that the curious name 'popeler' for the spoonbill may have been a mispronunciation of the Dutch *lepelaar*, which refers to the same bird. In German it is *Löffler* and in Frisian *leppelbek*, and all cases refer to the shape of bird's beak.

22 Gurney (1921: 123).

23 Cooke and Birkhead (2017).

24 Gurney (1921: 142).

25 Gurney (1834); Boorde's medical book – see Mullens and Swan (1917: 81).

26 Muffett's book was written about 1595, but not published until a century later. '*Venus*' sports and sparrows are referred to in chapter 11 of his book.

27 Some other examples: the fat or grease of a heron or crane placed in the ears for deafness, a young raven burnt to ashes, or the dung of a stork drunk in water for the falling sickness (epilepsy); a burnt cuckoo against 'the stone' – an extremely painful condition in which stones develop in the kidney, ureter or bladder. Finally, the ashes of a Little Owl taken in the throat will cure quinsy (a complication of tonsillitis). I recall my grandmother telling me how, as a child in the 1890s, she was seriously ill with a quinsy, and while not treated with owl ashes, she recovered when it finally 'broke' and she ejected a large string of pus from her mouth. Ugh!

28 Kioko et al. (2015).

Chapter 6: The New World of Science: Francis Willughby and John Ray Discover Birds

1 Farnley Hall itself lies on the north side of the Wharfe valley, about 20km north-west of Leeds, while the lake, Lake Tiny, lies in the adjacent Washburn valley.
2 Hill (1988); Lyles (1988).
3 The Farnley 'Ornithological Collection' contains patches of plumage of a wide range of birds shot on the estate mounted on sheets of paper. *The Ornithology of Francis Willughby* (Ray, 1678). Turner's bird paintings are now in Leeds City Art Gallery.
4 Shrubb (2013: 52); Hudson (*c.*1920); Muffett (1655: 93).
5 Ray (1678: 278); Gurney (1921); Lowe (1954: 147); Birkhead and Berkhoudt (2021).
6 Birkhead (2021).
7 I have presumed that it was Willughby rather than Ray who suggested overhauling the whole of natural history, on the basis of Willughby's more 'daring' mind (Birkhead, 2018).
8 Birkhead (2018).
9 The *Ornithology* was widely recognized as *the* ornithology text, by Newton (1896), Stresemann (1975) and others.
10 Birkhead (2018).
11 Grigson (2016): foreword by Juliet Clutton Brock.
12 Sossinka (1982).
13 Hume (2006).
14 Ray (1678: 153–4); the quote is from Strickland and Melville (1848: 22).
15 Hume (2006). An X-ray of this specimen's skull in 2018 revealed that it had been shot, raising questions about its provenance (Warnett et al., 2020).

16 Ray (1678: 152); Clusius (1605). For the Dodo phylogeny see Shapiro et al. (2002).

17 Evans (2000); Ray (1678: 245). As far as I can tell, the Northern Cardinal *Cardinalis cardinalis* does not feature in the Florentine Codex, but the Mexican House Finch *Carpodacus mexicanus* does, and Sahagún states that 'it is capable of domestication, it is teachable; it can be bred . . . They domesticate it. I domesticate it' (Sahagún, 1981: 48), suggesting that in the late 1500s it was kept as a cage bird by the Aztecs – as the Cardinal probably was too (Johnston, 2004).

18 Pierson (2001).

19 Ibid.

20 Varey (2001).

21 López-Ocón (2001).

22 Reeds (2002) – review of Varey et al. (2001) – and Freedberg (2002).

23 Reeds (2002).

24 Freedberg (2002: 247).

25 *Treasury of Medicinal Matters of New Spain or the History of Mexican Plants, Animals and Minerals* . . . Available at www.wdl.org/en/item/19340/ – as *Mexican treasure*, formally titled *Rerum medicarum Novae Hispaniae thesaurus, seu, Plantarum animalium mineralium Mexicanorum historia* (*Inventory of Medical Items from New Spain, or, History of Mexican Plants, Animals and Minerals*).

26 Ray (1678: 385).

27 Raven (1942).

28 Ray (1678: 392). As Norton (2019: 127) discovered, Hernández obtained much of his information on the quetzal from Sahagún's *Historia Universalis* (i.e. the Florentine Codex). In 1558 Sahagún had been instructed to document details of the Nahua way of life that would be useful in the maintenance of Christianity among the Nahua. He did this using indigenous

collaborators that included artists and scribes who interviewed Aztec elders. This was a sophisticated exercise that attempted to separate fact from fiction using the 'Aristotelian method of gathering factual information from knowledgeable persons and dependable manuscripts, with newer research methods that included the use of formalized questionnaires and peer review' (Haemig, 2018). The end result was an illustrated account of the natural history of Mexico written in both Nahuatl and Spanish. It provides a brief account and often an image – reminiscent of comic strips and not massively inferior to the woodcuts of Gessner and Belon in Europe – of 149 birds. It also reveals, for example, how the Nahua were using the wing feathers from wintering wildfowl to make writing quills before the Spanish conquest.

29 Norton (2019: 129).

30 Ibid.: 141.

31 Ibid. The flow of information is a complicated, multi-stranded, palimpsestic cascade of intelligence (see Varey (ed.), 2001) that flowed through the following: local Nahua people, educated (trilingual) Nahua who collaborated with the Spanish, Sahagún, Hernández (who lifted material from Sahagún), Recchi in the 1600s who paraphrased Hernández's account, Nieremberg (1635), who paraphrased Recchi's account, and John Ray, who copied Nieremberg's account word for word. The paper by Das and Lowe (2018) on decolonialism is relevant here.

32 Whitehead (1976: 411).

33 The original paintings are reproduced in the five volumes of *Brasil-Holandês* (*Dutch-Brazil*) (Anon., 1995).

34 King (2012: 4); Norton (2019: 121).

35 Norton (2019: 128 and 136). Codex Mendoza: https://iiif.bodleian.ox.ac.uk/iiif/viewer/2fea788e-2aa2-4f08-b6d9-

648c00486220#?c=0andm=0ands=0andcv=0andr=0andxywh=-3254%2C-464%2C13092%2C9265

36 Buono (2015). In 1539 the son-in-law of Moctezuma II commissioned a featherwork panel showing the Mass of St Gregory for Pope Paul III (who, two years previously had decreed that the indigenous people of Mexico had a soul and should not be enslaved). This panel, measuring 68 x 56 cm, whose background is a stunning, iridescent blue (possibly tanager feathers), reminded me of Peak District well dressings, created not with feathers but more modestly with flower petals and other vegetation. The history of well dressings is obscure, but I can easily imagine their being inspired by figurative featherworks brought to Europe, like the *Mass of St Gregory* (now in the Musée des Jacobins, in Auch, France).

37 From Amy Buono's review of King (2012): www.caareviews.org/reviews/2032/.

38 Sancho de la Hoz (1535).

39 The catfish that disrupts melanin metabolism is *Phractocephalus hemioliopterus* (Soares de Souza, 1851).

40 Teixeira (1985).

41 'It is better to have indigenous people killing wild animals [birds] to keep their original behaviour rather than a massive deforestation to increase the agriculture frontier.' Raphael Santos, pers. comm., 31 July 2020. See also Buono (2015: 181).

42 Norton (2012: 69).

43 Cited in Buono (2015: 183).

44 Norton (2012: 70).

45 Bujok (2004); Hesse (2010). Friedrich's father collected New World exotica and two of the three feather shields shown in the drawings made to orchestrate the procession are still in existence. This pageant celebrated Friedrich at the pinnacle of his power: on 24 January 1599 the Holy Roman emperor had agreed in the

Treaty of Prague that the territory of Württemberg would no longer be under imperial control, but would be under the absolute ownership of the dukes of Württemberg.

46 Buono (2015); and see Aztec processions in the Codex Mendoza (fol. 67r).

47 Drugs sent from Mexico to Europe included guaiacum (lignum vitae – a cure for syphilis – except it wasn't), balsam, jalap (a cathartic drug) and sassafras (for treating wounds/fever) and tobacco imported from New Spain. Also vanilla, tomato, corn, potatoes, chilli peppers and datura.

48 Arredondo and Bauer (2019).

49 Jacobs (2016: 21 and 18); Norton (2012: 69).

50 Norton (2012: 67).

51 Just as John Ogilvie struggled to appreciate the story accompanying the purchase of the headdress he wanted, Westerners often find indigenous stories difficult to understand because we expect a logical structure, and expect to take the story literally. In reality, many indigenous stories are like religious sermons, full of hidden meaning and metaphors designed to instil particular values and reinforce particular beliefs. There are also similarities between the stories told by indigenous peoples and those recounted by scientists in their seminars or publications, inasmuch that non-scientists – my parents, for example – needed to be guided through them sentence by sentence to obtain some level of understanding. It is exactly that kind of guidance I would like to help me understand Inuit stories. Indigenous 'storywork' is now a developing academic discipline (Archibald et al., 2019).

52 Ray (1678: preface, p. 5).

53 Hernández felt so indebted to the three Nahua artists – Pedro Vasquez, Anton Elias and Baltasar Elias – who accompanied him on his travels, that he bequeathed 60 ducats each to them

in his will. (Chabran and Varey 2001: 107). Sahagún names and acknowledges all those who helped him (Haemig, 2018: appendix). Note also that Hernández did not acknowledge that he took material from Sahagún's Florentine Codex. In today's science, a principal investigator typically acknowledges the help she or he may have received, but subsequent authors citing that work do not, since it is not (currently) part of academic etiquette.

54 Birkhead (2018). See Jacobs (2016) for details of Levaillant and Benson.

55 Birkhead (2018); Raven (1942).

56 Ray (1714).

57 Raven (1942: 465).

58 Ray (1714). Topsell (1972), or rather, Aldrovandi (1599), of whom his book *Fowles of Heauen* was a translation, anticipated some of this, and indeed I wonder whether Ray's statement in *The Wisdom of God* was inspired by what he had read in Aldrovandi.

59 Ray (1714).

60 Raven (1942: 466–7).

61 Ibid.: 457.

62 Haffer et al. (2014).

Chapter 7: Depending on Birds: Inconspicuous Consumption

1 Abacuk Pricket was one of several mutineers on the *Discovery*, who set captain Henry Hudson adrift with eight others (none of whom survived) in James Bay in June 1611. After taking control of the ship the mutineers called at Digges Island to harvest guillemots (murres) which is where Pricket found the 'round hills of stone'. The structures are similar to the cleits

on St Kilda used for the same purpose (http://archive.org/details/henryhudsonnavig27ashe/page/n15/mode/2up. My colleague Tony Gaston saw Pricket's 'hills of stone' intact while studying Brünnich's Guillemots (aka Thick-billed Murres) on Digges Island in the 1980s (Gaston et al. 1985).

2 Debes (1676: 134).

3 Otero et al. (2018).

4 Wylie (1987).

5 Debes (1676: 144–5).

6 Ibid.: 141. The Great Auk may once have bred on the Faroes.

7 Debes (2017: 290).

8 Ibid.; Avramov (2019). Note: 'magister' means 'scholar'.

9 Worm (1655); Debes (2017: 89).

10 Jens-Kjeld Jensen, pers. comm.

11 Didrik Sørensen (1802–65) lived on the island of Sandoy.

12 Williamson (1948: 146).

13 Debes (1676: 149–50).

14 Harris (2011: 168).

15 Vogt in Debes (2017: 308 n.1132).

16 Harris (2011); Tycho Anker-Nilssen, pers. comm.

17 Dunning et al. (2018).

18 Jensen (2010).

19 Olsen (2003); Joensen (1963). Our host on Skúvoy was Inga Tórarenni.

20 Dyck and Meltofte (1975); Jens-Kjeld Jensen, pers. comm.

21 Jens-Kjeld Jensen, pers. comm.

22 Williamson (1948: 157).

23 Vaughan (1998).

24 Frederickson et al. (2019).

25 Fielden (1872).

26 Landt (1810); Jensen (2010).

27 Jensen (2012: 25).
28 Birkhead (2016).
29 Birkhead (2014a); Nørrevang (1958).
30 Landt (1810). The cause of the fulmar's expansion is unknown. Ideas include: (i) more offal from whaling, (ii) a new genotype, or (iii) climate change – increase in areas of warm water.
31 Martin (1698: 57); Fisher (1952: 489).
32 Jensen (2010: 16).
33 Fisher (1952: 449).
34 Approximately 40 per cent of the fulmar chick's body mass at fledging is fat (Phillips and Hamer, 1999).
35 Herrmann et al. (2006); Fossádal et al. (2018).
36 Debes (1676: 376).
37 Ibid.: 109, 135; tongue cutting, see Birkhead (2008: 250).
38 Birkhead (2008: 257).
39 Grouw and Bloch (2015): the pied raven of the Faroes was partially leucistic, meaning an absence of melanin from certain feathers.
40 Macaulay (1764); Randall (2004); and see Seabird Harvest in the North Atlantic: www.atlanticseabirds.info/.
41 Martin (1698: 65).
42 Ibid.: 13.
43 Ibid.: 115; Nelson (1978: 286); Buchan (1727), cited in Chambers (2011). A Scottish stone was equivalent to 17lb 8oz, or 7.936kg (Wikipedia). Nelson (1978: 281) says '90,000 small seabirds' – for which I read puffins – were needed for 200 stone, and 'almost 300 gannets were needed to stuff one feather bed' Hutchinson (2014).
44 Martin (1698: 46).
45 www.atlanticseabirds.info/mykines.
46 See Mike Day's 2016 film *The Islands and the Whales*. Puffin information from Tycho Anker-Nilssen, pers. comm.

Chapter 8: The End of God in Birds: Darwin and Ornithology

1 Burkhardt et al. (1985 – vol. II: 11).

2 The Down House hummingbird display, a two-metre-long case, stuffed with birds collected in the 1840s, was donated to Down House in 1965 by the Lubbock family, descendants of Darwin's friend and neighbour Sir John Lubbock. See: http://darwin-online.org.uk/content/frameset?viewtype=text&itemID=A691&pageseq=1; Tree (1991).

3 Foster (1988).

4 Mabey (1986).

5 White (1789, 1813 edn: 491).

6 Dadswell (2003: 23).

7 Secord (2000).

8 King (2019: 5).

9 Jenyns was curate of Swaffham Bulbeck between 1827 and 1844; this is a quote from Jenyns's 'chapters of my life', cited by Wallace (2005: 66).

10 Barlow (1958).

11 Jenyns (1846: 139, 151, 184).

12 Wood (1862: 509).

13 Whittington-Egan (2014).

14 SPCK, established in 1698, is still going strong, and is currently Britain's leading publisher of educational Christian books.

15 Genesis 1: 21–2.

16 Syme (1823: 25).

17 Albin (1741); Holden (1875: 10).

18 Bechstein (1795) was translated into English and ran through numerous editions between its first publication and the late 1800s.

19 Wallace (1887).

20 Anon (1735).

21 Smith (1835: 5–6).

22 The Common Cuckoo and other brood parasites are exceptions – they seem to be able to function without their true parents to imprint upon: see Davies (2010: 155–8; 205–8).

23 Birkhead et al. (1990). When my son was at school and asked his father's profession, he referred to me as a 'professor of promiscuity'.

24 Barlow (1958).

25 Individual selection – beautifully explained in Richard Dawkins's *The Selfish Gene* (1976); The basics: within a species, individuals vary in various ways. If you were to measure the size of a particular bird species' beaks, you find some longer and deeper than others. Often, such variation has a genetic basis, so that beak size can be passed from parents to their offspring. Now imagine that a beak of a particular size means it is easier to eat seeds of a certain size: big beaks are better for big seeds, and small beaks for small seeds. Now imagine a drought that results in a lack of small seeds. The big-beaked birds survive on big seeds; the small-beaked birds, unable to open the big seeds, starve and die out. The next generation of birds then consists almost entirely of large-beaked birds. This is essentially what happened on the Galápagos Islands in the 1980s as a result of a drought, demonstrating at a stroke the reality of what Darwin had anticipated 150 years previously, following his own visit there in the 1830s. See Boag and Grant (1981); Lamichhaney et al. (2015, 2016).

26 Wallace (2005: 18).

27 Chassagnol (2010).

28 Chitty (1974).

29 Kingsley (1871).

30 Morris (1850–57, vol. V: 119).

31 Morris (1850–57, vol. III: 53). Note that pagination differs in the various editions of Morris; I have used the duodecimo edition.
32 Morris (1897: 213–15, 222).
33 Gosse (1907: 75, 139).
34 Ray (1678; 1714).
35 Rennie (1835: 314).
36 Jenner, cited in Davies (2015); Waterton (1871: 317).
37 Smith (1999).
38 Blackburn (1872, 1873).
39 Tree (1991: 214).
40 Holmes (2018).
41 Ibid.: 208, 232, 233. The natural-theological theme was further offset by placing Darwin's statue, created in 1885, in pride of place on the staircase. It is telling that Darwin's statue is in white (marble); and that of Owen, created later in 1897, is in black (bronze).
42 Gould's failure to obtain a Marvellous Spatuletail is described in Tree (1991: 162).
43 Smith (2006); for an alternative view see Frith (2016).

Chapter 9: A Dangerous Type of Bigamy: Killing Time

1 Gould (1861: vol. III, plate 172). The Blue-tailed Hummingbird Gould is referring to is what is now known as the Long-tailed Sylph.
2 Collar and Prys-Jones (2012) for this and the following on Hume.
3 Ibid. Jerdon's comment is from his Prospectus in his *Birds of India* regarding the proposed *The Fauna of British India, Including Ceylon and Burma.* See https://en.wikipedia.org/wiki/Thomas-C.-Jerdon.

4 Wood and Fyfe (1943).
5 Avery and Calaresu (2019).
6 Schulze-Hagen et al. (2003); Gessner (1555).
7 Schulze-Hagen et al. (2003).
8 Wagstaffe and Fidler (1955).
9 'Last of the great . . .' is from Garfield (2008). See also Knox (1993); Rasmussen and Collar (1999) on Meinertzhagen's misdemeanours.
10 Mearns and Mearns (1998).
11 Rothschild (1983).
12 Birkhead et al. (2014: 75–82).
13 Chapman (1933: 161).
14 Mearns and Mearns (1998: 21).
15 Cited in Mearns and Mearns (1998: 22).
16 Loss et al. (2012).
17 Nethersole-Thomson on Chance, cited in Cole and Trobe (2000: 35). Decline of the Red-backed Shrike in UK: since 1988 a few pairs have bred in England and the bird remains common on the Continent.
18 Letter from Beck to Clinton G. Abbott, cited in Abbott (1933).
19 Marshall (1912), cited in Collar and Prys-Jones (2012: 32). Hume married Mary Grindall in 1853, but almost nothing is known about her.
20 Marshall (1912).
21 Ibid.: 36.
22 Scientific name: *Calandrella acutirostris.*
23 See Mearns and Mearns (1998: 407). The Western Palearctic is a geographic region, and a 'tick' is how birders refer to new species they have seen; new either for a region or for a personal list.
24 The Bulo Burti Shrike was not the first new bird to be described from a live specimen: several much earlier ornithologists,

including W. Rothschild and P. L. Sclater in the 1800s, and Delacour and Jabouille in the 1930s, had done the same with certain species: Collar (1999); and for subspecies, see: Smith et al. (1991); Nguembock et al. (2008). The Blue-throated Hillstar is in the genus *Oreotrochilus*.

25 https://markavery.info/2018/11/15/ornithologists-kill-critically-endangered-hummingbirds/.

26 Porter, cited in Burton (2021).

27 Remsen (1995). This scholarly paper was unique in several respects, including the case *against* collecting. Remsen presents the following common objections to collecting, that it: (i) damages bird populations, (ii) is morally wrong, (iii) has enough specimens already, (iv) is unnecessary – a photograph and a blood sample for DNA is all that is needed, (v) depletes a country of its national heritage, and (vi) sets a bad example, in particular among local people. He then shows that none of these arguments stand up. Of all Remsen's arguments, his last (vi) seems the weakest, especially in the light of Richard Porter's earlier comment.

28 DiEuliis et al. (2016).

29 Dennis (2008).

30 Cole and Trobe (2000).

31 See: Holden and Holden (2017: 38) and www.theguardian.com/artanddesign/2017/sep/10/natural-selection-andy-peter-holden-artangel-newington-library-review. The two collectors were the multiple offenders Matthew Gonshaw and Colin Watson – the latter died, aged sixty-two, after falling from a tree in 2006 while climbing to the nest of a sparrowhawk, a protected species: see http://news.bbc.co.uk/1/hi/england/south_yorkshire/5294900.stm.

32 Walters (2005); Prys-Jones et al. (2019).

33 Johnson (2018). Edwin Rist was a twenty-year-old American musician fascinated by fly-tying and fishing flies.

34 Collar and Prys-Jones (2012).
35 Ibid: 22.

Chapter 10: Watching Birds: And Seeing the Light

1 Selous (1899, 1901). Note that Florence Merriam's interest in bird-watching predates his, but, unlike him, she never killed birds.
2 Selous (1899, 1901).
3 Some may argue that pioneers like Gilbert White had already provided a template, but, while there are some similarities, White, unlike Selous, did not sit and observe birds and think about why they were behaving in particular ways.
4 Sulloway (1996).
5 Howard (1920). A long list of more casual birders, as far back as Aristotle himself, and Frederick II in the thirteenth century, had alluded to the fact that many birds seemed to occupy a 'free hold'. As hinted at by Fisher (1966) and more explicitly by Kinlen (2018), Howard's ideas were probably not as original as he made out; but even so, his *Territory in Bird Life* developed the idea of territory and, by publicizing it so effectively, stimulated a huge amount of research on this fundamental aspect of avian life.
6 Birkhead et al. (2014).
7 Ibid. Claud Ticehurst, appointed editor in 1931, was the main offender.
8 Alexander (1915). For Emma Turner see: Parry and Greenwood (2020).
9 Witherby et al.'s five-volume *Handbook of British Birds* (1938–1941); Lockley (1947); Alexander (1915).

10 Niemann (2013). Interestingly, as far as we can tell, no German PoWs in Britain used birdwatching as a distraction (D. Niemann, pers. comm.). Fisher (1940); Buxton (1950); Moss (2004).

11 Ogilvie et al. (2007).

12 Hartley (1954); see also Wallace (2004: 123).

13 Summers-Smith et al. (1954).

14 Hartley (1954). Peter Harold Trahair Hartley (1909–94); obtained a first-class honours in zoology in 1933 at the University of London. He was a military man, active in conservation, and from 1970 to 1975 Archdeacon of Suffolk. He was a colleague of David Lack's in Oxford, with whom he shared similar views. As if to emphasize how easily tribal elitism can distort our view of birds, David Lack was told that the first draft of his Prefatory Note to *Enjoying Ornithology* gave 'two undesirable and unintended impressions . . . that all ornithologists should do research and that it is a waste of time to watch birds in other ways, and that ornithologists that do research are superior to that who do not'. In the final version Lack (1965) apologizes and modifies his text accordingly.

15 Tinbergen (1953); Lorenz (1952); Armstrong (1955).

16 Moss (2004).

17 Cocker (2001).

18 Ibid.

19 A survey of *British Birds* readers in 2010 showed that 97% of 882 respondents (21% of those surveyed) were male (C. Spooner, pers. comm.). For changes in gender among those interested in birds, see Huang et al. (2020); Australian birding – Penny Olsen, pers. comm. To document the increasing number of women involved in ornithology and biology, my colleague Bob Montgomerie, at Queen's University in Ontario, looked at the numbers

of papers in *The Auk*, *The Condor* and *The American Naturalist* in which women were first authors. His results show a dramatic increase starting in the 1970s that was probably a result of expanding educational opportunities and emancipation (R. Montgomerie, pers. comm.). Several websites celebrate the role of women in ornithology: https://matthewhalley.wordpress.com/2018/01/17/female-pioneers-of-ornithology/; www.discoverwildlife.com/people/the-wonder-women-of-ornithology/; https://americanornithology.org/the-invisible-women/.

The proportion of women as authors of scientific papers increased from 14% in 1945 to 27% in 2010, but this was accompanied by 'an increase in the gender disparities in productivity and impact' (Huang et al., 2020).

Similar 'niche separation' of bird interests is apparent elsewhere. In North America, the most basic level was served by *Audubon* magazine and *Natural History*; the next tier was *Birding* for hard-core birders, then *Bird Banding/Journal of Field Ornithology*, *Canadian Field Naturalist* for birders engaged in surveys, followed by *Wilson Bulletin*, *Condor* and *Auk* for the serious, independent bird researchers, followed by the international concept journals.

20 Medawar (1979); in *The Blind Watchmaker* (1986), Richard Dawkins argues that biology is infinitely more complicated than physics. This is despite the fact that physicists are often considered smarter than biologists: www.discovermagazine.com/the-sciences/physicists-vs-biologists-cognitive-smackdown. Lack (1965) also proffers some guidance – now dated, but not useless – on securing a career in birds.

21 Lau et al. (2020).

22 These examples are all from the first volume of *British Birds* in 1907/8.

23 Side Hunt: see Birkhead et al. (2014).

24 The first student was Tim Birch and the second was Keith Clarkson, both of whom went on to have successful careers in conservation.

25 Snetsinger (2003). Hornbuckle: see www.shanghaibirding.com/hornbuckle/. The top bird lister is constantly shifting, because of new species and new taxonomy. Twitching: see Cocker (2001); Moss (2004); Mynott (2009).

26 www.10000birds.com/how-many-birders-are-there-really-updated.htm.

27 eBird: https://ebird.org/science; www.icarus.mpg.de/4158/icarus-global.

Chapter 11: A Boom in Bird Studies: Behaviour, Evolution and Ecology

1 Stresemann (1975); Schulze-Hagen and Kaiser (2020).

2 Heinroth (1911).

3 Heinroth (1924–1933). Schulze-Hagen and Kaiser (2020): *Die Vogel-WG*. The title requires some explanation: *WG* refers to *Wohngemeinschaft*, which means a shared flat or apartment, a term that perfectly describes how the Heinroths lived with their birds. (Dr Gabriele Kaiser is archivist in the Department of Manuscripts in Berlin's Staatsbibliothek.) Their book, published in 2020, resulted in a minor media sensation in Germany – the Heinroths finally receiving some of the recognition so long deserved. I eagerly await the book's translation into English.

4 Schulze-Hagen and Kaiser (2020).

5 Ibid.; Schulze-Hagen and Birkhead (2015).

6 Birkhead et al. (2014).

7 Karl Schulze-Hagen, Heinroth's biographer, told me how Heinroth protected and sheltered several Jewish friends during the Nazi regime.

8 Stresemann (1975: 348); Schulze-Hagen and Kaiser (2020).

9 Kruuk (2003); Burkhardt (2005).

10 Burkhardt (2005).

11 Julian Huxley had previously identified three of these; Tinbergen, as he readily and modestly admitted, merely added development. But it was Tinbergen's paper that set the course for the study of behaviour – see Birkhead et al. (2014: 281).

12 Tinbergen (1953).

13 Such was the separation between research groups – Tinbergen's Animal Behaviour Research Group and Lack's Edward Grey Institute of Field Ornithology, with its focus on ecology – their ideas failed to connect. Towards the ends of their lives, Tinbergen and Lack became close, sharing confidences about their illnesses – Lack's cancer (non-Hodgkin's lymphoma) and Tinbergen's depression (K. Schulze-Hagen, pers. comm., from their correspondence).

14 Birkhead et al. (2014); the ornithologist David Snow (pers. comm.) told me that Wynne-Edwards had once told him that he considered his idea as important as Darwin's.

15 In the early 1970s my fiancée and I had a camping holiday in Switzerland. On meeting a colleague at the Swiss Ornithological Station at Sempach, it was suggested that if we would allow an eastern European visitor to accompany us they would cover all our costs. With only limited funds, we agreed, but it quickly became clear why they wanted rid of their visitor. The first evening, I pointed to the far end of the tent and told our guest that was where he should sleep. 'No!' he said, pointing to my partner and indicating that he wanted to sleep next to

her. 'No!' I said, and like Trivers's male pigeon, I positioned myself between my fiancée and our unwelcome passenger, only to find in the morning that he had somehow managed to position himself next to her, but fortunately still in his sleeping bag.

16 Birkhead and Monaghan (2010); it was not only Oxford, but also Harvard, where Trivers and E. O. Wilson were based, Imperial College London, where W. D. (Bill) Hamilton worked, and John Maynard Smith in Sussex.

17 Segerstråle (2000).

18 Nelson (2013); John Krebs and Nick Davies's textbook *Behavioural Ecology* (1978 – and subsequent editions) was, and still is, the 'bible'.

19 Birkhead and Montgomerie (2020).

20 Bob May, pers. comm. to N. B. Davies and J. R. Krebs, independently.

21 Beautifully described by Nick Davies in his book *Cuckoo: Cheating by Nature* (2015).

22 Bonhote (1907); Krebs and Davies (1978); Stacey and Koenig (2008).

23 Darwin (1871).

24 Selous (1901).

25 Birkhead (1992).

26 Birkhead et al. (2014: viii).

Chapter 12: Ghost of the Great Auk: Third Mass Extinction

1 In May 1907, Newton started to suffer from dropsy – fluid retention – 'his first real illness in seventy-eight years'. Writing to his friend John Harvie-Brown, he said that while he wasn't

in pain, he fully expected the dropsy to carry him off – which it did. He also said: 'I wish I could have lived to tell "The Story of the Gare-fowl" [Great Auk] and "The Bustard and Britain", for which I have laid in a vast stock of material, but perhaps someone else may be found to use it . . .' (Wollaston 1921). See also Birkhead and Gallivan (2012); Urry (2021).

2 For three excellent ornithological biographies, see Hale (2016); McGhie (2017) and Milsom (2020).

3 Waterton (1871: 411).

4 Newton (1869).

5 Grieve (1885); see also Birkhead and Gallivan (2012) and Pálsson (in press). Why garefowl? In Icelandic, the Great Auk's name *geir-fugl*, 'spear bird', is presumably a reference to its powerful beak. The Inuit name was apparently *isarukitsok*, meaning 'little wing', referring, obviously, to the Great Auk's disproportionately tiny wings. 'Little Wing' was also the title of one of Jimi Hendrix's most remarkable and beautiful songs, which I like to think was about the Great Auk. *Isarukitsok* is from Clements Markham (1875).

6 Fuller (1999).

7 Burkhardt et al. (1985–, vol. XIX), letter to E. R. Lankester, 22 March [1871]; www.darwinproject.ac.uk/letter/DCP-LETT-7612.xml. Sympathy as unmanly: see Endersby (2009).

8 Cowles (2013).

9 Morris (1867).

10 Knocker: see http://medicalgentlemen.co.uk/patients-and-diseases/knocker. In a breeding season lasting 110 days, twenty-five boats, each with two guns, killed between them 975 birds each day, resulting in a seasonal total of some 107,250 birds.

11 Morris (1868). Prince Christian and seabirds eggs – from Barnes-Lawrence's diary. See also Hickling (2021) for legal aspects of the new Acts.

12 £3.19s. is equivalent to about £500 today (E. Boardman, pers. comm. and see https://bridlingtonpriory.co.uk/wp-content/uploads/2020/01/Resource-Booklet-Rev-Henry-Barnes-Lawrence-and-the-Sea-Birds-Preservation-Act.pdf.

13 Newton (1896: 398); Morris (1897); Vaughan (1998: 145). The increase in guillemot numbers, if true, would have been remarkable, since seabird populations rarely change so rapidly.

14 The Puritans of the Massachusetts Bay Colony in 1641 introduced the first statutory measures protecting animals from mistreatment: Watson (2014).

15 Ritvo (1989: 128).

16 Ibid.

17 Ibid.: 138.

18 http://news.bbc.co.uk/local/humberside/hi/people_and_places/nature/newsid_9383000/9383787.stm.

19 Thoreau (1853–4: 66). This 'spy-glass' was an army telescope. Binoculars became available much later, in Britain not until 1896. Before then, opera glasses were often used, as promoted by Florence Merriam in her 1889 best-selling book *Birds Through an Opera Glass.*

20 Allen (1886). 'Leprosy' is Max Nicholson's description in his 1926 book, *Birds in England.* To see Britain's sluggishness in this regard, see Moss (2004) and Birkhead et al. (2014: 168–9).

21 Newton (1861). The idea of bird censuses did not originate with Newton; as with all great ideas, others have usually suggested something similar before. The great American ornithologist Alexander Wilson, for example, had conducted surveys of the birds in the Pennsylvania botanical gardens in 1811.

22 Hollom field guide: Peterson et al. (1954). For Hollom, see Porter (2015). *The Most Offending Soul Alive* is the title of

Heimann's (1998) biography of Harrisson. The main threats to bird populations across the world are, in descending order: (i) habitat loss (e.g. deforestation); (ii) invasive species (e.g. house mice on ocean islands that kill albatross adults and chicks); (iii) domestic cats that across the globe kill millions of birds each day; (iv) hunting, as in countries like Egypt, where millions of migrant birds are killed and eaten; (v) trade in birds, that just as with the Great Auk, follows a 'rarity-feedback model', a clunky term that simply but sadly means that the rarer a species becomes, the greater the demand for it and the higher price people are prepared to pay. Overall, 1,469 bird species (13 per cent of the total) are globally threatened and a further 1,017 are near-threatened. Together, these represent a fifth of all bird species. Overall, 40 per cent of all bird species are in decline, 44 per cent are stable, only 7 per cent are increasing and the status of the remaining 9 per cent is unknown (www.birdlife.org/wp-content/uploads/2021/02/SOWB2018-en.pdf). The picture is a grim one. It is hard to imagine that the Atlantic Puffin and the Black-legged Kittiwake are both considered vulnerable to extinction. Since 1500, some 161 bird species have disappeared: an unprecedented rate of extinction. The last known Hawaiian Poo-uli (Black-faced Honeycreeper) died in captivity in 2004.

23 Birkhead (2014b).

24 The guillemot's pyriform-shaped egg is inherently more stable and less likely to be dislodged than any other shape, especially on a sloping ledge. This is because, with its long straight edge, there's more egg (and more friction) in contact with the substrate. Two earlier explanations, both now debunked, are that this shape allows the egg to spin on its axis or roll in an arc. Other possibilities are that this shape is more resistant to

damage or keeps the blunt end out of the muck on dirty ledges, but stability seems to be the key (Birkhead et al., 2018).

25 Meade et al. (2012).

26 Birkhead (2016).

27 Votier et al. (2009).

28 Charmantier and Gienapp (2013).

29 The situation on Shetland is most clearly seen on Fair Isle: www.fairislebirdobs.co.uk/seabird_research.html. Isle of May: Ashbrook et al. (2008). West coast of North America: Piatt et al. (2020). See also Fayet et al. (2021) for the effect of changes of food distribution on Atlantic Puffins.

Epilogue

1 'The bleedin' obvious' was a phrase made popular by John Cleese in the 1970s cult TV series *Fawlty Towers*. Soon after the publication of Gladwell's book in 2000, he, I and two other authors took part in Radio 4's *Start the Week*, hosted by the irascible Jeremy Paxman. As quickly became clear, Paxman also considered the idea of *The Tipping Point* to be rather obvious.

2 Vaccination did, however, have its roots in indigenous, non-scientific cultures (see Ridley, 2020).

3 Compare Birkhead (2018) and (2021).

4 If you want to know what things could be like if we do not address climate change, watch Barry Hines, *Threads*: www.youtube.com/watch?v=s_s8CrRN76Mandlist=PL13xVFVD-3WSWHqvA4DOK77r_1FXmhgf5; Lopez (2019: 46).

5 Birkhead and Nettleship (1995).

6 Thomas (1983); Fowles (1979).

7 Snow (1959); see also New Networks for Nature as an example of a movement whose aim is to integrate different views of nature: www.newnetworksfornature.org.uk/.

8 It is now well established that interactions with nature generally increase our sense of well-being; a review of the growing interest in human–nature interactions is provided by Soga and Gaston (2020).

Bibliography

Abbott, C. G. (1933). Closing history of the Guadalupe Caracara. *Condor* 35: 10–14.

Albin, E. (1741). *A Natural History of English Song-birds*. Bettesworth and Co., London.

Aldrovandi, U. (1599–1603). *Ornithologiae hoc est de avibus historiae*. Bologna.

Alexander (1915). A practical study of bird oecology. *British Birds* 8: 184–92.

Allen, E. G. (1951). The history of American ornithology before Audubon. *Transactions of the American Philosophical Society* 41: 386–591.

Allen, J. A. (1886). The present wholesale destruction of bird-life in the United States. *Science* 7: 191–5.

Anon. (1735). *The Bird Fancier's Recreation*. Smith, London.

Anon. (1995). *Brasil-Holandês* [*Dutch-Brazil*]. Editorial Index, Rio de Janeiro.

Anon. (no date). Rev. H. F. Barnes-Lawrence of Bridlington Priory and the Sea Birds Preservation Act 1869, https://bridlingtonpriory.co.uk/rev-henry-barnes-lawrence-and-the-1869-sea-birds-preservation-act/

Archibald, J. et al. (2019) (eds.). *Decolonizing Research: Indigenous Storywork as Methodology*. Zed, London.

Aristotle (1936) *Marvellous Things Heard*. In W. S. Hett (ed.), *Aristotle: Minor Works*. Harvard University Press, Cambridge, MA.

——— (1937). *Parts of Animals*, ed. A. L. Peck and E. L. Forster. Harvard University Press, Cambridge, MA.

——— (1943). *Generation of Animals*, ed. T. E. Page. Harvard University Press, Cambridge, MA.

——— (1965). *History of Animals*, ed. A. L. Peck. Harvard University Press, Cambridge, MA.

Armstrong, E. A. (1955). *The Wren*. Collins, London.

Arredondo, J. M. and Bauer, R. (2019) (eds.). *Translating Nature: Cross Cultural Hierarchies of Early Modern Science*. University of Pennsylvania Press, Philadelphia.

Ashbrook, K. et al. (2008). Hitting the buffers: conspecific aggression undermines benefits of colonial breeding under adverse conditions. *Biology Letters* 4: 630–33.

Avery, V. and Calaresu, M. (2019) (eds.). *Feast and Fast*. Cambridge University Press, Cambridge.

Avramov, I. (2019). Letters and questionnaires: the correspondence of Henry Oldenburg and the early Royal Society of London's Inquiries for Natural History. Chapter 5 in P. Findlen (ed.), *Empires of Knowledge: Scientific Networks in the Early Modern World*. Routledge, London.

Bahn, P. G. (2016). *Images of the Ice Age*. Oxford University Press, Oxford.

Bailleul-LeSuer, R. (2013). *Between Heaven and Earth: Birds in Ancient Egypt*. Oriental Institute of the University of Chicago, Chicago.

Baker, E. C. S. (1921). *The Game Birds of India, Burma and Ceylon*, vol. 2. John Bale, Sons and Danielsson, London.

Barlow, N. (1958). *The Autobiography of Charles Darwin*. Collins, London.

Bates, C. (2011). George Turberville and the painful art of falconry. *English Literary Renaissance* 41: 403–28.

Bechstein, J. M. (1795). *Natural History of Cage Birds*. Groombridge, London.

Belon, P. (1555). *Histoire de la nature des Oyseaux*. Prévost, Paris.

Bert, E. (1619). *An Approved Treatise on Hawks and Hawking.* Printed by T[homas] S[nodham] for Richard Moore, London.

Birds Preservation Act (1869). Heritage Open Days, https://bridlingtonpriory.co.uk/rev-henry-barnes-lawrence-and-the-1869-sea-birds-preservation-act/.

Birkhead, T. R. (1992). *The Magpies.* Poyser, London.

——— (2008). *The Wisdom of Birds.* Bloomsbury, London.

——— (2014a). An academic life: researching and teaching animal behaviour. *Animal Behaviour* 91: 5–10.

——— (2014b). Guillemots on Skomer: the value of long-term population studies. *Natur Cymru* 2014: 10–15.

——— (2016). Changes in the numbers of Common Guillemots on Skomer since the 1930s. *British Birds* 109: 651–9.

——— (2018). *The Wonderful Mr Willughby.* Bloomsbury, London.

——— (2021). Cracking the mystery. *BBC Wildlife*, March 2021: 72–6.

Birkhead, T. R., Atkin, L. and Møller, A. P. (1988). Copulation behaviour in birds. *Behaviour* 101: 101–38.

Birkhead, T. R. and Berkhoudt, H. (2021). Francis Willughby at Sevenhuis in June 1663. Pp. 78–88 in A. van de Haar and A. Schulte Nordholt (eds.), *Figurations animalières à travers les textes et l'image en Europe: Du Moyen Âge à nos jours.* Brill, Leiden.

Birkhead, T. R. and Gallivan, P. (2012). Alfred Newton's contribution to ornithology: a conservative quest for facts rather than grand theories. *Ibis* 154: 887–905.

Birkhead, T. R. and Lessells, C. M. (1988). Copulation behaviour of the osprey *Pandion haliaetus. Animal Behaviour* 36: 1672–82.

Birkhead, T. R. and Monaghan, P. (2010). Ingenious ideas: the history of behavioral ecology. Pp. 3–15 in D. F. Westneat and C. Fox (eds.), *Evolutionary Behavioral Ecology*, Oxford: Oxford University Press.

Birkhead, T. R. and Montgomerie, R. (2020). Three decades of sperm competition in birds. *Philosophical Transactions of the Royal Society* 375, https://doi.org/10.1098/rstb.2020.0208.

Birkhead, T. R. and Nettleship, D. N. (1995). Arctic fox influence on a seabird community in Labrador: a natural experiment. *Wilson Journal of Ornithology* 107: 397–412.

Birkhead, T. R., Thompson, J. E. and Montgomerie, R. (2018). The pyriform egg of the Common Murre (*Uria aalge*) is more stable on sloping surfaces. *Auk* 135: 1020–32.

Birkhead, T. R., Wimpenny, J. and Montgomerie, R. (2014). *Ten Thousand Birds: Ornithology Since Darwin*. Princeton University Press, Princeton.

Birkhead, T. R., Wishart, G. J. and Biggins, J. D. (1995). Sperm precedence in the domestic fowl. *Proceedings of the Royal Society of London Ser B* 261: 285–92.

Birkhead, T. R. et al. (1990). Extra-pair paternity and brood parasitism in wild Zebra Finches *Taeniopygia guttata*, revealed by DNA fingerprinting. *Behavioural Ecology and Sociobiology* 27: 315–24.

Blackburn, J. H. (1872). Cuckoo and pipit. *Nature* 5: 383.

——— (1873). Cuckoos. *Nature* 9: 123.

Bloch, D. (2012). Beak tax to control predatory birds in the Faroe Islands. *Archives of Natural History* 39: 126–35.

Bloch, H. (2005). Animal fables, the Bayeux Tapestry, and the making of the Anglo-Norman world. *Poetica* 37: 285–309.

Boag, P. T. and Grant, P. R. (1981). Intense natural selection in a population of Darwin's finches (Geospizinae) in the Galápagos. *Science* 214: 82–5.

Bock, W. J. (2015). Evolutionary morphology of the woodpeckers (Picidae). *Denisia 36, zugleich Katalogue des oberösterreichiscen Landesmuseum*, Neue Serie 164: 37–54.

Bonhote, J. L. (1907). Four birds in a Long-tailed Tit's nest. *British Birds* 1: 62.

Brehm, A. E. (1867). *Tierleben*. Bibliographischen Instituts, Hildburghausen. Bresciani, E. (1980). *Kom madig 1977e 1978: Le Pitturi Murali del Cenotafio di Alessandre Magno*. Giardini Editori, Pisa.

Breuil, H. and Burkitt, M. C. (1929). *Rock Paintings of Southern Andalusia*. Oxford University Press. Oxford.

Briggs, S. (2014). Catherine caged: birds in the margins of the Hours of Catherine of Cleves. *Bowdoin Journal of Art* 2014: 1–18.

Brock, R. (2004). Aristotle on sperm competition in birds. *Classical Quarterly* 54: 277–8.

Broderick, A. H. (1963). *The Abbé Breuil, Prehistorian: A Biography*. Hutchinson, London.

Buffon, G. L. (1770–83). *Histoire Naturelle des Oiseaux*. Imprimerie Royale, Paris. Trans. W. Smellie (1792–3). Strahan and Cadell, London.

Bujok, E. (2004). *Neue Welten in europäischen Sammlungen. Africana und Americana in Kunstkammern bis 1670*. Reimer, Berlin.

Buono, A. J. (2015). 'Their treasures are the feathers of birds': Tupinambá featherwork and the image of America. Pp. 178–88 in A. Russo, G. Wolf, and D. Fane (eds.), *Images Take Flight: Feather Art in Mexico and Europe 1400–1700*. Hirmer, Munich.

Burkhardt, F. et al. (1985–) (eds.). *The Correspondence of Charles Darwin*. Cambridge University Press, Cambridge.

Burkhardt, R. W. Jr. (2005). *Patterns of Behavior: Konrad Lorenz, Niko Tinbergen, and the Founding of Ethology*. University of Chicago Press, Chicago.

Burton, J. (2021). The killing fields of wildlife, https://johnandrewsson.wordpress.com/2021/03/30/the-killing-fields-of-wildlife/.

Buxton, J. (1950). *The Redstart.* Collins, London.

Cade, T. (1982). *Falcons of the World.* Collins, London.

Canby, J. V. (2002). Falconry (hawking) in Hittite lands. *Journal of Near Eastern Studies* 61: 161–201.

Chabran, R. and Varey, S. (2001). Entr'acte. Pp. 105–8 in Varey et al. (2001). Chabran and Weiner (eds.), *Searching for the Secrets of Nature.*

Chambers, B. (2011) (ed.). *Rewriting St Kilda: New Views on Old Ideas.* The Islands Book Trust, Lewis.

Chapman, A. (1930). *Memories of Fourscore Years Less Two, 1851–1929.* Gurney and Jackson, Edinburgh.

Chapman, A. and Buck, W. J. (1910). *Unexplored Spain.* Arnold, London.

Chapman, F. M. (1933). *Autobiography of a Bird-Lover.* Appleton-Century, New York.

Charmantier, A. and Gienapp, P. (2013). Climate change and timing of avian breeding and migration: evolutionary versus plastic changes. *Evolutionary Applications*, https://doi.org/10.1111/eva.12126.

Chassagnol, A. (2010). Darwin in wonderland: evolution, involution and natural selection in *The Water-Babies* (1863). *Miranda*, DOI: https://doi.org/10.4000/miranda.376.

Chitty, S. (1974). *The Beast and the Monk: A Life of Charles Kingsley.* Hodder and Stoughton, London.

Cleland, J. (1607). *Institution of a Young Noble Man.* Ioseph Barnes, Oxford.

Clottes, J. (2016). *What is Palaeolithic Art? Cave Paintings and the Dawn of Human Creativity.* University of Chicago Press, Chicago.

Clusius, C. (1605). *Exoticorum libri decem.* Ex Officinâ Plantinianâ Raphelengii.

Clutton-Brock, J. (1989). Review of Houlihan and Goodman. *Antiquity* 63: 386–7.

Cocker, M. (2001). *Tales of a Tribe.* Jonathan Cape, London.

Cole, A. C. and Trobe, W. M. (2000). *The Egg Collectors of Great Britain and Ireland.* Peregrine Books, Leeds.

Cole, F. J. (1944). *The History of Comparative Anatomy.* Macmillan, London.

Collar, N. J. (1999). New species, high standards and the case of *Laniarius liberatus. Ibis* 141: 358–67.

Collar, N. J. and Prys-Jones, R. P. (2012). Pioneer of Asian ornithology: Allan Octavian Hume. *BirdingASIA* 17: 17–43.

Cooke, F. and Birkhead, T. R. (2017). The identity of the bird known locally in sixteenth- and seventeenth-century Norfolk, United Kingdom, as the spowe. *Archives of Natural History* 44: 118–21.

Cornellá, M. M. (2003–5). Willoughby Verner y la Laguna de la Janda. *Archia* 3–5: 225–30.

Cowles, H. M. (2013). A Victorian extinction: Alfred Newton and the evolution of animal protection. *British Journal for the History of Science* 46: 695–714.

Cox, N. (1686). *The Gentleman's Recreation.* Blome, London.

Cummins, J. (1988). *The Hound and the Hawk.* Weidenfeld and Nicolson, London.

Cuvier, G. (1817). *Essay on the Theory of the Earth.* Blackwood, Edinburgh.

Dadswell, T. (2003). *The Selborne Pioneer: Gilbert White as Naturalist and Scientist: A Re-examination.* Ashgate, Abingdon.

Dakin, R. and Montgomerie, R. (2013). Eye for an eyespot: how iridescent plumage ocelli influence peacock mating success. *Behavioral Ecology* 24: 1048–57.

Darwin, C. (1871). *Descent of Man and Selection in Relation to Sex.* Murray, London.

Das, S. and Lowe, M. (2018). Nature read in black and white: decolonial approaches to interpreting natural history collections. *Journal of Natural Science Collection* 6: 4–14, www.natsca.org/article/2509.

Davies, N. B. (2010). *Cuckoos, Cowbirds and Other Cheats.* Poyser, London.

———(2015). *Cuckoo: Cheating by Nature.* Bloomsbury, London.

Davies, N. de G. (1900–1901). *The Mastaba of Ptahhetep and Akhethetep at Saqqareh.* Egypt Exploration Fund, London.

Dawkins, R. (1976). *The Selfish Gene.* Oxford University Press, Oxford.

———(1986). *The Blind Watchmaker.* Norton, New York.

de Juana, E. and Garcia, E. (2015); *The Birds of the Iberian Peninsula.* Bloomsbury, London.

Debes, L. (1676). *Færoæ, and Færoa reserata: That is a Description of the Islands Inhabitants of Foeroe: Being Seventeen Islands Subject to the King of Denmark*, William Iles, London.

———(2017) *A Description of Foeroe, 1676*, ed. N. B. Vogt. Stidin, Tórshavn.

Dennis, R. (2008). *A Life of Ospreys.* Whittles, Caithness.

Díaz-Andreu, M. (2013). The roots of the first Cambridge textbooks on European prehistory: an analysis of Miles Burkitt's formative trips to Spain and France. *Complutum* 24: 109–20.

DiEuliis, D. et al. (2016). Opinion: Specimen collections should have a much bigger role in infectious disease research and response. *Proceedings of the National Academy of Sciences* 113: 4–7.

Douglas, J. (1714). The natural history and description of the phoenicopterus or flamingo; with two views of the head, and three of the tongue, of that beautiful and uncommon bird. *Philosophical Transactions of the Royal Society* 29: 523–41.

Dunning, J. et al. (2018). Photoluminescence in the bill of the Atlantic Puffin *Fratercula arctica. Bird Study* 65: 570–73.

Dyck, J. and Meltofte, H. (1975). The guillemot *Uria aalge* population of the Faroes 1972. *Dansk Ornitologisk Forenings Tidsskrift* 69: 55–64.

Endersby, J. (2009), Sympathetic science: Charles Darwin, Joseph Hooker, and the passions of Victorian naturalists. *Victorian Studies* 51: 299–320.

Evans, S. T. (2000). Aztec royal pleasure parks: conspicuous consumption and elite status rivalry. *Studies in the History of Gardens and Designed Landscapes* 20: 206–28.

Fayet, A. L. et al. (2021). Local prey shortages drive foraging costs and breeding success in a declining seabird, the Atlantic Puffin. *Journal of Animal Ecology* 90: 1152–64.

Fielden, H. (1872). Birds of the Faeroe Islands. *Zoologist* 1872: 3277–94.

Fisher, J. (1940). *Watching Birds.* Penguin, London.

——— (1952). *The Fulmar.* Collins, London.

——— (1966). *The Shell Bird Book.* Ebury Press/Michael Joseph, London.

Fossádal, M. E., Grand, M. and Gaini, S. (2018). *Chlamydophila psittaci* pneumonia associated to the exposure to fulmar birds (*Fulmarus glacialis*) in the Faroe Islands. *Infectious Diseases* 50: 817–21.

Foster, P. G. M. (1988). *Gilbert White and His Records: A Scientific Biography.* Helm, Bromley.

Fowles, J. (1979). *The Tree.* Aurum Press, London.

Fox, N. (1995). *Understanding the Bird of Prey.* Hancock House, Blaine.

Frederickson, M. et al. (2019). Quantifying the relative importance of hunting and oiling on Brünnich's Guillemots in the north-west Atlantic. *Polar Research* 38: 3378, http://dx.doi.org/10.33265/polar.v38.3378.

Freedberg, D. (2002). *The Eye of the Lynx.* University of Chicago Press, Chicago.

Freeman, G. E. and Salvin, F. H. (1859). *Falconry: Its Claims, History, and Practice.* Longman, London.

Frith, C. B. (2016). *Charles Darwin's Life with Birds*. Oxford University Press, Oxford.

Fuller, E. (1999). *The Great Auk*. Privately printed.

Gardiner, A. H. (1928). *Catalogue of the Egyptian Hieroglyphic Printing Type, from Matrices Owned by Dr Alan Gardiner*. Oxford University Press, Oxford.

———(1929). Additions to the new hieroglyphic fount 1928. *Journal of Egyptian Archaeology* 15: 95.

———(1957). *Egyptian Grammar*, 3rd edn. Griffith Institute, Oxford.

Garfield, B. (2008). *The Meinertzhagen Mystery: The Life and Legend of a Colossal Fraud*. Potomac, Washington DC.

Gaston, A. J. et al. (1985). A Natural History of Digges Sound. *Canadian Wildlife Service Report* 46: 1–62.

Gessner, C. (1555). *Historium animalium liber III, qui est de Avium Natura*. Froschauer, Zurich.

———(1560). *Icones avium omnium*. Froschauer, Zurich.

Ghosh, S. K. (2015). Human cadaveric dissection: a historical account from ancient Greece to the modern era. *Anatomy and Cell Biology* 48: 153–69.

Gibson, G. (2005). *The Bedside Book of Birds*. Bloomsbury, London.

Gladwell, M. (2000). *The Tipping Point*. Little, Brown, New York.

Gosse, E. (1907). *Father and Son*. Heinemann, London.

Gould, J. (1838). *A Monograph of the Trogonidae, or Family of Trogons*. London.

———(1849–61). *A Monograph of the Trochilidae* (5 vols.). Taylor and Francis, London.

Grieve, S. (1885). *The Great Auk or Garefowl (Alca impennis): Its History, Archaeology and Remains*. T. Jack, Edinburgh.

Grigson, C. (2016). *Menagerie: The History of Exotic Animals in England*. Oxford University Press, Oxford.

Grouw, H. and Bloch, D. (2015). History of the extant museum specimens of the Faroese white-speckled raven. *Archives of Natural History* 42: 23–38.

Gurney, D. (1834). Extracts from the Household and Privy Purse Accounts of the Lestranges of Hunstanton, from A.D. 1519 to A.D. 1578. *The Gentleman's Magazine*, September 1834: 269.

Gurney, J. H. (1921). *Early Annals of Ornithology*. Witherby, London.

Haemig, P. D. (2018). A comparison of contributions from the Aztec cities of Tlatelolco and Tenochtitlan to the bird chapter of the Florentine Codex. *Huitzil Revista Mexicana de Ornithologiá* 19: 540–68.

Haffer, J., Hudde, H. and Hillcoat, B. (2014). The development of ornithology and species knowledge in central Europe. *Bonn Zoological Bulletin Supplement* 59: 1–116.

Hale, W. G. (2016). *Sacred Ibis: The Ornithology of Canon Henry Baker Tristram DD, FRS*. Sacristy, Durham.

Harris, M. P. (2011). *Puffins*. Poyser, London.

Harting, J. E. (1871). *The Birds of Shakespeare*. Van Voorst, London.

——— (1901). *A Handbook of British Birds*. Nimmo, London.

Hartley, P. H. T. (1954). Back garden ornithology. *Bird Study* 1: 18–27.

Harwood, D. (1928). *Love for Animals and How It Developed in Great Britain*. Columbia University Press, New York.

Haskins, C. H. (1921). The 'De Arte Venandi cum Avibus' of the Emperor Frederick II. *English Historical Review* 36: 334–55.

Heimann, Judith M. (1998). *The Most Offending Soul Alive: Tom Harrisson and His Remarkable Life*. University of Hawaii Press, Honolulu.

Heinroth, O. (1911). Beitrag zur Biologie, namentlich Ethologie und Psychologie der Anatiden. *Ber V. Internat OrnithologKongr Berlin 1910*: 589–702.

Heinroth, O. and Heinroth, M. (1924–33). *Die Vögel Mitteleuropas in allen Lebens- und Entwicklungsstufen photographisch aufgenommen und in ihrem Seelenleben bei der Aufzucht vom Ei ab beobachtet [The Birds of Central Europe – Photographed in All Stages of Life and Development and Observed in Their Mental Life During Rearing from the Egg]* (4 vols.). Hugo Behrmühler Verlag, Berlin.

Herrmann, B. et al. (2006). *Chlamydophila psittaci* in Fulmars, the Faroe Islands. *Emerging Infectious Diseases* 12: 330–32.

Hesse, S. (2010). Die Neue Welt in Stuttgart: Die Kunstkammer Herzog Friedrichs I und der Auufzug zum Ringrennen a 25 Februar 1599. Pp. 139–66 in J. Kremer, S. Lorenz and P. Rückert (eds.), *Hofkultur um 1600. Die Hofmusik Herzog Friedrichs I. von Württemberg und ihr kulturelles Umfeld.* Thorbecke, Ostfildern.

Hickling, J. (2021). The *vera causa* of endangered species legislation: Alfred Newton and the Wild Bird Preservation Acts, 1869–1894. *Journal of the History of Biology*, https://doi.org/10.1007/s10739-021-09633-w.

Hill, D. (1988). *Turner's Birds.* Phaidon, London.

Holden, A. and Holden, P. (2017). *Natural Selection.* Artangel, London.

Holden, C. F. (1875). *Holden's Book on Birds.* New-York Bird-Store, Boston.

Holmes, J. (2018). *The Pre-Raphaelites and Science.* Yale University Press, New Haven.

Holmes, R. (2010). The Royal Society's lost women scientists. *Guardian*, 21 November 2010, www.theguardian.com/science/2010/nov/21/royal-society-lost-women-scientists.

Houlihan, P. (1986). *Birds of Ancient Egypt.* Aris and Phillips, Warminster.

Howard, H. E. (1920). *Territory in Bird Life.* Murray, London.

Huang, J. et al. (2020). Historical comparison of gender inequality in scientific careers across countries and disciplines. *Proceedings of the National Academy of Sciences* 117: 4609–16.

Hudson, W. H. (*c.*1920). *The Book of a Naturalist.* Nelson, London.

Hume, J. P. (2006). The history of the Dodo *Raphus cucullatus* and the penguin of Mauritius. *Historical Biology* 18: 69–93.

Hutchinson, E. (1974). Attitudes towards nature in Medieval England: the Alphonso and bird psalters. *Isis* 65: 5–37.

Hutchinson, R. (2014). *St Kilda: A People's History.* Birlinn, Edinburgh.

Huxley, T. H. (1870). *Lay Sermons, Addresses and Reviews.* Macmillan, London.

Ikram, S. (2015) (ed.). *Divine Creatures; Animal Mummies in Ancient Egypt.* The American University in Cairo, Cairo.

Isaacson, W. (2017). *Leonardo da Vinci.* Simon and Schuster, London.

Jackson, C. E. (1993). *Great Bird Paintings of the World,* vol. 1. Antique Collectors' Club, Woodbridge.

Jacob, G. (1718). *The Compleat Sportsman.* Nutt and Gosling, London.

Jacobs, N. J. (2016). *Birders of Africa: History of a Network.* Yale University Press, New Haven.

James, M. R. (1925). An English medieval sketch-book, no. 1916 in the Pepysian Library, Magdalene College Cambridge. *The Thirteenth Volume of the Walpole Society* 13: 1–17.

Jensen, J.-K. (2010). *Puffin Fowling: A Fowling Day on Nólsoy.* Sjónband, Tórshavn.

——— (2012). *The Fulmar on the Faroe Islands* (in Faroese). Ritograk, Tórshavn.

Jenyns, L. (1846). *Observations in Natural History.* Van Voorst, London.

Joensen, A. H. (1963). Ynglefuglene på Skúvoy, Færøerne, deres udbredelse og antal. *Dansk Ornitologisk Forenings Tidsskrift* 57: 1–18.

Joffe, S. N. and Buchanan, V. (2016). The Andreas Vesalius woodblocks: a four-hundred-year journey from creation to destruction. *Acta Medico-Historia Adriatica* 14: 347–72.

Johns, C. A. (1862). *British Birds in Their Haunts*. SPCK, London.

Johnson, K. W. (2018). *The Feather Thief: Beauty, Obsession and the Natural History Heist of the Century*. Penguin, London.

Johnston, D. W. (2004). The earliest illustrations and descriptions of the cardinal. *Banisteria* 24: 3–7.

King, A. (2019). *The Divine in the Commonplace: Reverent Natural History and the Novel in Britain*. Cambridge University Press, Cambridge.

King, H. (2012). *Peruvian Featherworks*. Yale University Press, New Haven and London.

King, M. L. (1980). Book-lined cells: women and humanism in early Italian Renaissance. Pp. 66–90 in P. H. Labalme (ed.), *Beyond Their Sex: Learned Women of the European Past*. New York University Press, New York.

Kingsley, C. (1871). *At Last: A Christmas in the West Indies*. Macmillan, London.

Kinlen, L. (2018). Eliot Howard's 'law of territory' in birds: the influence of Charles Moffatt and Edmund Selous. *Archives of Natural History* 45: 54–68.

Kioko, J., Smith, D. and Kiffner, C. (2015). Uses of birds for ethno medicine among the Maasai people in Monduli district, northern Tanzania. *International Journal of Ethnobiology and Ethnomedicine* 1: 1–13.

Kleczkowska, K. (2015). Bird communication in ancient Greek and Roman thought. *Maska* 28. 95–106.

Knox, A. (1993). Richard Meinertzhagen: a case of fraud examined. *Ibis* 135: 320–25.

Krebs, J. R. and Davies, N. B. (1978). *Behavioural Ecology*. Blackwell, Oxford.

Krüger, T. et al. (2020). Persecution and statutory protection have driven Rook *Corvus frugilegus* population dynamics over the past 120 years in NW Germany. *Journal of Ornithology* 16: 569–84.

Kruuk, H. (2003). *Niko's Nature: The Life of Niko Tinbergen and His Science of Animal Behaviour*. Oxford University Press, Oxford.

Lack, D. (1965). *Enjoying Ornithology*. Methuen, London.

——— (1968). *Ecological Adaptations for Breeding in Birds*. Methuen, London.

Lamichhaney, S. et al. (2015). Evolution of Darwin's finches and their beaks revealed by genome sequencing. *Nature* 518: 371–5.

——— (2016). A beak size locus in Darwin's finches facilitated character displacement during a drought. *Science* 352: 470–74.

Landauer, W. (1961). *Hatchability of Chicken Eggs as Influenced by Environment and Heredity*. Storrs Agricultural Experiment Station, Storrs, CT.

Landt, G. (1810). *A description of the Feroe Islands*. Longman, Hurst, Rees and Orme, London.

Lau, J. K. L. et al. (2020). Shared striatal activity in decisions to satisfy curiosity and hunger at the risk of electric shocks. *Nature Human Behaviour* 4: 531–43 (2020), https://doi.org/10.1038/s41562-020-0848-3.

Lazarich, M., Ramos-Gil, A. and González-Pérez, J. L. (2019). Prehistoric bird watching in southern Iberia? The rock art of Tajo de las Figuras reconsidered. *Environmental Archaeology*, 24: 387–99, https://doi.org/10.1080/14614103.2018.1563372.

Lecky, W. H. L. (1913 edn). *History of European Morals from Augustus to Charlemagne*. Longmans, New York.

Leroi, A. M. (2014). *The Lagoon: How Aristotle Invented Science.* Bloomsbury, London.

Lockley, A. (2013). *Island Child: My Life on Skokholm with R. M. Lockley.* Gwasg Carreg Gwalch, Llanrwst.

Lockley, R. M. (1947). *Letters from Skokholm.* Dent, London.

Lones, T. E. (1912). *Aristotle's Researches in Natural Science.* West, Newman and Co., London.

Lopez, B. (2019). *Horizon.* Bodley Head, London.

López-Ocón, L. (2001). The circulation of the work of Hernández in nineteenth-century Spain. Pp. 183–93 in Varey et al. (2001).

Lorenz, K. (1952). *King Solomon's Ring.* Methuen, London.

Loss, S. R., Will, T. and Marra, P. P. (2012). The impact of free-ranging domestic cats on wildlife in the United States. *Nature Communications* 4: 1396, https://doi.org/10.1038/ncomms2380.

Lovegrove, R. (2007). *Silent Fields: The Long Decline of a Nation's Wildlife.* Oxford University Press, Oxford.

Lowe, F. (1954). *The Heron.* Collins, London.

Lyles, A. (1988). *Turner and Natural History at Farnley Park.* Tate Gallery, London.

Mabey, R. (1986). *Gilbert White: A Biography of the Author of the Natural History of Selborne.* Profile, London.

Macaulay, K. (1764). *The History of St Kilda.* Becket and de Hondt, London.

MacGillivray, W. (1837–52). *A History of British Birds, Indigenous and Migratory* (5 vols.). Scott, Webster and Geary, London.

Macias, S. (2011). *Mosaicos de Mértola: a arte bizantina no ocidente mediterrânico.* Câmara Municipal de Mértola, Mértola.

Macpherson, H. A. (1896). *A History of Fowling.* Douglas, Edinburgh.

Magnus, O. (1555). *Historia om de nordiska folken.* Stockholm.

Manley, D. and Ree, P. (2001). *Henry Salt: Artist, Traveller, Diplomat, Egyptologist.* Libri, London.

Markham, C. (1875). Papers on the Greenland Eskimos. In: *A Selection of Papers on Arctic Geography and Ethnology*. Murray, London.

Markham, G. (1621). *Hungers Prevention: or The Whole Art of Fowling By Water and Land*. Holme and Langley, London.

Marshall, C. H. T. (1912). Mr. Hume's work as an ornithologist. *India*, 2 August 1912: 57–8.

Martin, G. T. (1981). *The Sacred Animal Necropolis at North Saqqara*. Egypt Exploration Society, London.

Martin, M. (1698). *A Late Voyage to St Kilda*. Brown and Godwin, London.

McCouat, P. (2015). Lost masterpieces of ancient Egyptian art from the Nebamun tomb-chapel *Journal of Art in Society*, www.artinsociety.com/lost-masterpieces-of-ancient-egyptian-art-from-the-nebamun-tomb-chapel.html.

McGhie, H. A. (2017). *Henry Dresser and Victorian Ornithology: Birds, Books and Business*. University of Manchester Press, Manchester.

Meade, J. et al. (2012). The population increase of common guillemots *Uria aalge* on Skomer Island is explained by intrinsic demographic properties. *Journal of Avian Biology* 44: 55–61.

Mearns, B. and Mearns, R. (1998). *The Bird Collectors*. Poyser, London.

Medawar, P. J. (1979). *Advice to a Young Scientist*. Basic Books, New York.

Medawar, P. J. and Medawar, J. S. (1983). *Aristotle to Zoos*. Harvard University Press, Cambridge, MA.

Merriam, F. (1889). *Birds Through an Opera Glass*. Houghton Mifflin, New York.

Miller, J. (1991). *Charles II*. Weidenfeld and Nicolson, London.

Milsom, T. (2020). *Henry Seehbom's Ornithology*. Privately published.

Molina, V. (1913). Arqueología y prehistoria de la provincial de Cádiz en Lebria y Medinasidonia. *Boletín de la Real Academia de la Historia*: 554–62.

Morris, F. O. (1850–57). *A History of British Birds* (8 vols.). Groombridge, London.

——— (1867). Letter to *The Times*, 3 April 1867.

——— (1868). Letter to *The Times*, 25 August 1868.

Morris, M. C. F. (1897). *Francis Orpen Morris: A Memoir.* Nimmo, London.

Moser, S. (2020). *Painting Antiquity: Ancient Egypt in the Art of Lawrence Alma-Tadema, Edward Poynter and Edwin Long.* Oxford University Press, Oxford.

Moss, S. (2004). *A Bird in the Bush: A Social History of Birdwatching.* Aurum Press, London.

Muffett, T. (1655). *Health's Improvement.* Thomson, London.

Mullens, W. H. and Swan, H. K. (1917). *A Bibliography of British Ornithology.* Macmillan, London.

Mynott, J. (2009). *Birdscapes: Birds in Our Imagination and Experience.* Princeton University Press, Princeton.

——— (2018). *Birds in the Ancient World: Winged Words.* Oxford University Press, Oxford.

Nash, T. (1633). *Quaternio, the foure-fold Way to a happie Life.* Dawson, London.

Naumann, J. A and Naumann, J. F. (1820–60). *Naturgeschichte der Vögel Deutschlands.* E. Fleischer, Leipzig.

Nelson, B. (1978). *The Gannet.* Poyser, Berkhamsted.

Nelson, J. B. (2013). *On the Rocks.* Langford, Langtoft.

Newmyer S. T. (2011). *Animals in Greek and Roman Thought: A Sourcebook.* Routledge, New York.

Newton, A. (1861). On the possibility of taking an ornithological census. *Ibis* 3: 190–96.

——— (1869). The zoological aspect of game laws: Address to the British Association, Section D, August 1868. Repr. 1893 in Society for the Protection of Birds, *Third Annual Report*, Appendix: 24–31.

———(1896). *A Dictionary of Birds*. Black, London.

Newton, I. and Olsen, P. (1990) (eds.). *Birds of Prey*. Murdoch, Sydney.

Nguembock, B. et al. (2008). Phylogeny of *Lanairius*: molecular data reveal *L. liberatus* synonymous with *L. erlangeri* and 'plumage coloration' as unreliable morphological characters for defining species and species groups. *Molecular Phylogenetics and Evolution* 48: 396–407.

Nice, M. M. (1979). *Research is a Passion with Me*. Consolidated Amethyst Publications, Toronto.

Nicholson, E. M. (1926). *Birds in England*. Chapman and Hall, London.

Niemann, D. (2013). *Birds in a Cage*. Short, London.

Nieremberg, J. E. (1635). *Historia naturae, maxime peregrinae, libris XVI*. Antwerp.

Nørrevang, A. (1958). On the breeding biology of the guillemot (*Uria aalge*). *Dansk Ornitologisk Forenings Tidsskrift* 53: 48-74.

Norton, M. (2012). Going to the birds. Pp. 53–83 in P. Findlen (ed.), *Early Modern Things: Objects and Their Histories, 1500–1800*. Routledge, London.

———(2019). The quetzal takes flight: microhistory, Mesoamerican knowledge, and early modern natural history. Pp. 119–47 in J. M. Arredondo and R. Bauer (eds.), *Translating Nature: Cross-Cultural Histories of Early Modern Science*. University of Pennsylvania Press, Philadelphia.

Oggins, R. (2004). *The Kings and Their Hawks: Falconry in Medieval England*. Yale University Press, New Haven.

Ogilvie, M., Ferguson-Lees, J. and Chandler, R. (2007). A history of British Birds. *British Birds* 100: 3–15.

Olsen, I. (2003). Bestandsudviklingen af ynglefuglene på Skúvoy, Farøerne 1961–2001 [Population development of breeding birds on Skúvoy, Faroe Islands, 1961–2001]. *Dansk Ornitologisk Forenings Tidsskrift* 97: 199–209.

Otero, X. L. et al. (2018). Seabird colonies as important global drivers in the nitrogen and phosphorus cycles. *Nature Communications* 9: 246, https://doi.org/10.1038/s41467-017-02446-8.

Oviedo Gonzalo Fernández de Valdés (1526). *La natural y hystoria de la Indias*. Toledo.

Owen-Crocker, G. R. (2005). Squawk talk: commentary by birds in the Bayeux Tapestry. *Anglo-Saxon England* 34: 237–54.

Pálsson, G. (in press). *An Awkward Extinction*. Princeton University Press, Princeton.

Pantsov, A. and Levine, S. I. (2013). *Mao: The Real Story*. Simon and Schuster, London.

Parkinson, R. (2008). *The Painted Tomb-Chapel of Nebamun*. British Museum Press, London.

Parry, J. and. Greenwood, J. (2020). *Emma Turner: A Life Looking at Birds*. Norfolk and Norwich Naturalists' Society, Norwich.

Perelló, E. H. (1988). Abate H. Breuil y coronel W. Verner: textos sobre la cueva de la Pileta. Pp. 173–81 in *Actas del Congreso Internacional 'El Estrecho de Gibraltar' Ceuta, 1987*. Universidad Nacional de Educación a Distancia, Madrid.

Peterson, R. T., Mountford, G. and Hollom, P. A. D. (1954). *A Field Guide to the Birds of Britain and Europe*. Collins, London.

Phillips, R. A. and Hamer, K. C. (1999). Lipid reserves, fasting capability and the evolution of nestling obesity in Procellariiform seabirds. *Proceedings of the Royal Society B* 266: 1329–34.

Piatt, J. et al. (2020). Extreme mortality and reproductive failure of Common Murres resulting from the northeast Pacific marine heatwave of 2014–2016. *PLoS ONE* 15(1): e0226087, https://doi.org/10.1371/journal.pone.0226087.

Pierson, P. O. (2001). Philip II: Imperial obligations and scientific vision. Pp. 11–18 in Varey et al. (2001).

Piso, G. and Marcgrave, G. (1648). *Historia naturalis Brasiliae.* Hackium, Leiden.

Pizzari, T. et al. (2003). Sophisticated sperm allocation in male fowl. *Nature* 426: 70–74.

Pliny (1885). *Naturalis Historia*, Book X: *The Natural History of Birds.* Taylor and Francis, London.

Porter, R. (2015). Obituary: PAD Hollom. *Sandgrouse* 37: 111–12.

Pricket, A. (1610). A larger discourse of the same voyage, and the successe thereof. Pp. 98–136 in G. M. Asher (ed.) (2016), *Henry Hudson the Navigator: The Original Documents in Which His Career is Recorded.* Hakluyt Society, London.

Prys-Jones, R., Adams, M. and Russell, D. G. (2019). Theft from the Natural History Museum's bird collection – what can we learn? *Alauda* 87: 73–82.

Randall, J. (2004) (ed.). *Traditions of Seabird Fowling in the North Atlantic Region.* Islands Book Trust, Inverness.

Ransome, A. (1947). *Great Northern?* Jonathan Cape, London.

Rasmussen, P. C. and Collar, N. J. (1999). Major specimen fraud in the Forest Owlet *Heteroglaux* (*Athene* auct.) *blewitti. Ibis* 141: 11–21.

Raven, C. E. (1942). *John Ray, Naturalist: His Life and Works.* Cambridge University Press, Cambridge.

Ray, J. (1678). *The Ornithology of Francis Willughby.* Martyn, London.

——— (1714). *The Wisdom of God Manifested in the Works of the Creation.* Originally published 1691. Samuel Smith, London.

Réaumur, M. de (1750). *The Art of Hatching and Bringing up Domestick Fowls of all Kinds at any Time of the Year, either by means of the heat of Hot-beds, or that of Common Fire.* Royal Academy of Sciences, Paris.

Reeds, K. (2002). What the Nahua knew: review of Varey et al. (2001). *Nature* 416: 369–70.

Remsen, J. V. (1995). The importance of continued collecting of bird specimens to ornithology and bird conservation. *Bird Conservation International* 5: 145–80.

Rennie, J. (1835). *The Faculties of Birds*. Knight, under the superintendence of the Society for the Diffusion of Useful Knowledge, London.

Richardson, B. W. (1885). Vesalius, and the Birth of Anatomy. In *The Asclepiad: A Book of Original Research and Observation in the Science, Art, and Literature of Medicine, Preventive and Curative* (8 vols.), vol. II. Longmans, Green, London.

Ridley, M. (2020). *How Innovation Works*. Fourth Estate, London.

Ritvo, H. (1989). *The Animal Estate*. Harvard University Press, Cambridge, MA.

Robinson, G. (2003). *The Sinews of Falconry: From Earliest Times Until the Epoch of the 1950–65 Pesticide Crisis*. Privately printed.

Rothschild, M. (1983). *Dear Lord Rothschild: Birds, Butterflies and History*. Hutchinson, London.

Sahagún (1981). *Florentine Codex*, ed. C. E. Dibble and J. O. Anderson. Monographs of the School of American Research and the Museum of New Mexico 14, part XII.

Sancho de la Hoz, P. (1535). Relación para S. M. de lo sucedido en la conquista. Pp. 117–79 in H. H. Urteaga (1938) (ed.), *Los cronistas de la conquista*. Desclée de Brouwer, Paris.

Schulze-Hagen, K. and Birkhead, T. R. (2015). The ethology and life history of birds: the forgotten contribution of Oskar, Magdalena and Katharina Heinroth. *Journal of Ornithology* 156: 9–18.

Schulze-Hagen, K. and Kaiser, G. (2020). *Die Vogel-WG: Die Heinroths ihre 1000 Vogelund*. Knesebeck, Munich.

Schulze-Hagen, K. et al. (2003). Avian taxidermy in Europe from the Middle Ages to the Renaissance. *Journal of Ornithology* 144: 459–78.

Secord, J. A. (2000). *Victorian Sensation.* University of Chicago Press, Chicago.

Segerstråle, U. (2000). *Defenders of the Truth: The Battle for Science in the Sociology Debate and Beyond.* Oxford University Press, Oxford.

Selous, E. (1899). An observational diary of the habits of nightjars, mostly of a sitting pair. *Zoologist* 3: 388–402.

——— (1901). *Bird Watching.* J. M. Dent, London.

Shapiro, B. et al. (2002). Flight of the Dodo. *Science* 295: 1683.

Shrubb, M. (2013). *Feasting, Fowling and Feathers: A History of the Exploitation of Wild Birds.* Poyser, London.

Sick, H. (1993). *Birds in Brazil: A Natural History*, trans. W. Belton. Princeton University Press, Princeton.

Singer, C. (1931). *A Short History of Biology.* Clarendon Press, Oxford.

Smellie, W. (1790). *The Philosophy of Natural History.* The Heirs of Charles Elliot, Edinburgh.

Smith, E. F. G. et al. (1991). A new species of shrike (Laniidae: *Laniarius*) from Somalia, verified by DNA sequence data from the only known individual. *Ibis* 133: 227–35.

Smith, F. (1835). *The Canary: Its Varieties, Management and Breeding.* Groombridge, London.

Smith, J. (1999). The cuckoo's contested history. *Trends in Ecology and Evolution* 14: 415.

——— (2006). *Charles Darwin and Victorian Visual Culture.* Cambridge University Press, Cambridge.

Smith, J. E. H. (2018). The ibis and the crocodile: Napoleon's Egyptian campaign and evolutionary theory in France, 1801–1835. *Republic of Letters* 6: 1–20.

Smith, P. (2007). On toucans and hornbills: readings in early modern ornithology from Belon to Buffon. Pp. 75–119 in K. A. E. Enekel and M. S. Smith (eds.), *Early Modern Zoology: The Construction of Animals in Science, Literature and the Visual Arts.* Brill, Leiden.

Snetsinger, P. (2003). *Birding on Borrowed Time*. American Birding Association, Colorado Springs.

Snow, C. P. (1959). *The Two Cultures and the Scientific Revolution*. Cambridge University Press, Cambridge.

Soares de Souza, G. (1851). *Tratado descriptivo do Brasil em 1587*. Instituto Histórico e Geográfico Brasileiro, Rio de Janeiro.

Soeteboom, H. (1648) (ed.). *Schipper Willem van West-Zanen's Reys na de Oost-Indien, A°1602 &c*... H. Soeteboom, Amsterdam.

Soga, M. and Gaston, K. J. (2020). The ecology of human–nature interactions. *Proceedings of the Royal Society B* 287: 20191882, http://dx.doi.org/10.1098/rspb.2019.1882.

Sossinka, R. (1982). Domestication in birds. *Avian Biology* 6: 373–403.

Stacey, P. and Koenig, W. (2008) (eds.). *Cooperative Breeding in Birds: Long Term Studies and Behaviour*. Cambridge University Press, Cambridge.

Stresemann, E. (1975). *Ornithology: From Aristotle to the Present*. Harvard University Press, Cambridge, MA. Published originally in 1951 as *Entwicklung Der Ornithologie von Aristotles bis zur Gegenwart*.

Strickland, H. E. and Melville, A. G. (1848). *The Dodo and Its Kindred*. Reeve, Benham and Reeve. London.

Strutt, J. (1842). *The Sports and Pastimes of the People of England*. Bohn, London.

Sulloway, F. (1996). *Born to Rebel: Birth Order, Family Dynamics, and Creative Lives*. Pantheon, New York.

Summers-Smith, D., Yeates, G. K. and Scott, R. R. (1954). Scientific ornithology. *Bird Study* 1: 71–2.

Syme, P. (1823). *A Treatise on British Song-birds*. Anderson, Edinburgh.

Teixeira, D. M. (1985). Plumagens aberrantes em psittacidae neotropicais. *Revista Brasileira de Biologia* 45: 143–8.

Thomas, K. (1983). *Man and the Natural World: Changing Attitudes in England, 1500–1800*. Allen Lane, London.

Thoreau, H. (1853–4). *The Writings of Henry David Thoreau: Journal*, vol. V: *March 5, 1853–November 30, 1853*. Princeton University Press, Princeton.

——— (1862). Walking. *The Atlantic*, June 1862.

Tinbergen, N. (1953). *The Herring Gull's World*. Collins, London.

——— (1960). The evolution of behavior in gulls. *Scientific American* 203: 118–33.

——— (1963). Aims and methods of Ethology. *Zeitschrift für Tierpsychologie* 20: 410–33.

Topsell, E. (1972). *The Fowles of Heauen or History of Birdes*. University of Texas Press, Austin.

Traves, G. (2006). *Flamborough: A Major Fishing Station*. Privately printed.

Tree, I. (1991). *The Bird Man: A Biography of John Gould*. Ebury Press, London.

Turberville, G. (1575). *The Booke of Faulconrie or Hauking*. Facsimile, New York, 1969.

Urry, A. (2021). Hearsay, gossip, misapprehension: Alfred Newton's secondhand histories of extinction. *Archives of Natural History* 48: 244–62.

Usick, P. (2007). Review of Bednarski's 'Holding Egypt . . .'. *Journal of Egyptian Archaeology* 93: 308–10.

Vansleb, M. (1678). *The Present State of Egypt, or, A New Relation of a Late Voyage into That Kingdom Performed in the Years 1672 and 1673*. John Starkey, London.

Varey, S. (2001). Francisco Hernández, Renaissance man. Pp. 33–40 in Varey et al. (2001).

Varey, S. (2001) (ed.). *The Mexican Treasury: The Writings of Dr Francisco Hernández*. Stanford University Press, Stanford.

Varey, S., Chabran, R. and Weiner, D. B. (2001). *Searching for the Secrets of Nature: The Life and Works of Dr Francisco Hernández*. Stanford University Press, Stanford.

Vaughan, R. (1998). *Seabird City: Guide to the Breeding Seabirds of the Flamborough Headland*. Smith Settle, Otley.

Vaurie, C. (1971). Birds in the prayer-book of Bonne of Luxembourg. *Bulletin of the Metropolitan Museum of Art, New York* 29: 279–81.

Venturi, A. (1904). *Storia dell'arte Italiana*. Hoepli, Milan.

Verner, W. W. C. (1909). *My Life Among the Wild Birds in Spain*. J. Bale, Sons and Danielsson Limited, London.

——— (1911). Letters from wilder Spain. *Saturday Review* 112: 360–61, 395–7, 422–4, 458–9, 483–4, 518–19.

——— (1914). Prehistoric man in southern Spain. *Country Life* 911: 901–4; 914: 41–5; and 916: 114–18.

Villing, A. et al. (2013). *Naukratis: Greeks in Egypt*. British Museum Press, London.

von den Driesch, A. et al. (2005). Mummified, deified and buried at Hermopolis Magna – the sacred ibis birds from Tuna el-Gebel, Middle Egypt. *Ägypten und Levante / Egypt and the Levant* 15: 203–44.

Votier, S. et al. (2009). Changes in the timing of egg-laying of a colonial seabird in relation to population size and environmental conditions. *Marine Ecology Progress Series* 393: 225–33.

Wagstaffe, R. and Fidler, J. H. (1955). *Preservation of Natural History Specimens*. Riverside Press, New York.

Wallace, D. I. M. (2004). *Beguiled by Birds*. Helm, London.

Wallace, L. (2005). *Leonard Jenyns: Darwin's Lifelong Friend*. Bath Royal Literary and Scientific Institution, Bath.

Wallace, R. L. (1887). *British Cage Birds*. Upcott Gill, London.

Walters, M. P. (2005). My life with eggs. *Zoologische Mededelingen* 79: 5–18.

Warnett, J. M. et al. (2020). The Oxford Dodo. Seeing more than ever before: X-ray micro-CT scanning, specimen acquisition and provenance. *Historical Biology*, https://doi.org/10.1080/08912963.2020.1782396.

Wasef, S. et al. (2019). Mitogenomic diversity in sacred ibis mummies sheds light on early Egyptian practices. *PLoS ONE* 14(11): e0223964, https://doi.org/10.1371/journal.pone.0223964.

Waterton, C. (1871). *Essays on Natural History*. Warne, London.

Watson, R. N. (2014). Protestant animals: Puritan sects and English animal-protection sentiments, 1550–1650. *English Literary History* 81: 1111–48.

West, M. and King, A. P. (1990). Mozart's starling. *American Scientist* 78: 106–14.

Whitaker, J. (2002) (ed.). *The Natural History Diaries of Willoughby Verner: Being an Account of His Natural History Expeditions, 1867–1890*. Peregrine Press, Leeds.

White, G. (1789). *The Natural History and Antiquities of Selborne*. B. White and Son, London.

Whitehead, P. J. P. (1976: 411). The original drawings for the *Historia naturalis Brasiliae* of Piso and Marcgrave (1648). *Journal of the Society for the Bibliography of Natural History* 7: 409–22.

Whittington-Egan, R. (2014). *The Natural History Man: A Life of the Reverend J. G. Wood*. Cappella Archive, Malvern.

Wilde, W. R. (1840). *Narrative of a Voyage to Madeira, Teneriffe and Along the Shores of the Mediterranean*. Curry, Dublin.

Wilkin, S. (1835). *Sir Thomas Browne's Works*. Pickering, London.

Wilkinson, J. G. (1878). *Manners and Customs of the Ancient Egyptians*. Murray, London.

Willemsen, C. A. (1943). *Die Falkenjagd. Bilder aus dem Falkenbuch Kaiser Friedrichs II*. Insel, Leipzig.

Williams, G. C. (1966). *Adaptation and Natural Selection*. Princeton University Press, Princeton.

Williamson, K. (1948). *The Atlantic Islands: A Study of the Faeroe Life and Scene*. Collins, London.

Witherby, H. F. et al. (1938–1941). *The Handbook of British Birds* (5 vols.). Witherby, London.

Wollaston, A. F. R. (1921). *A Life of Alfred Newton, Professor of Comparative Anatomy, Cambridge University, 1866–1907*. Murray, London.

Wood, C. A. and Fyfe, F. M. (1943). *The Art of Falconry, Being the De Arte Venandi cum Avibus of Frederick II of Hohenstaufen*. Stanford University Press, Stanford.

Wood, J. G. (1862). *The Illustrated Natural History: Birds*. Routledge, London.

Worm, O. (1655). *Museum Wormianum*. Elzevier, Leiden.

Wylie, J. (1987). *The Faroe Islands: Interpretations of History*. University of Kentucky Press, Lexington.

Wynne-Edwards, V. C. (1962). *Animal Dispersion in Relation to Social Behaviour*. Oliver and Boyd, Edinburgh.

Yapp, W. B. (1979). The birds of English medieval manuscripts. *Journal of Medieval History* 5: 315–48.

———(1981). *Birds in Medieval Manuscripts*. British Library, London.

———(1982). Birds in captivity in the Middle Ages. *Archives of Natural History* 10: 479–500.

———(1983). The illustrations of birds in the Vatican manuscript of *De art venandi cum avibus* of Frederick II. *Annals of Science* 40: 597–634.

———(1987). Animals in medieval art: the Bayeux Tapestry as an example. *Journal of Medieval History* 13: 15–73.

Yarrell, W. (1843). *A History of British Birds* (3 vols.). Van Voorst, London.

Index

Page references in *italics* indicate images.